D0032226

Praise for *Diary of a Young Naturalist*:

'Dara is an extraordinary voice and vision: brave, poetic, ethical, lyrical, strong enough to have made him heard and admired from a young age'
– Robert Macfarlane

'An extraordinary diary . . . it's a powerful pitch for why the school curriculum needs to be wilded and a reminder of the value of neurodiversity in literature'
– *The Times*, a *Times* Nature Book of the Year 2020

'A torrent of pure, unmediated fervour . . . an extraordinarily accomplished work for any writer, let alone one who is still a teenager . . . This is writing at its wild and unruly best'
– Dr Rachel Clarke, *The Lancet*

'It's a diary but essentially timeless. It's about enduring, it's about passion, beauty and connection. It's really, really special'
– Chris Packham

'Wise, lyrical and well-researched . . . [Dara's] candid enthusiasm, his powers of observation, his passion for nature – all are being rediscovered by a world population forced to stop short and take stock'
– *The Irish Times*

'An exceptional new voice. Dara McAnulty celebrates nature in such a fresh way and illuminates our understanding of autism'
– Martha Kearney, chair of the Baillie Gifford Prize 2020

'*Diary of a Young Naturalist* is not only one of the finest pieces of modern nature writing produced on this island in recent years, McAnulty is one of our best young writers in any genre'
– *Irish Independent*

'Dara is only 16 and autistic, and is already on
his way to becoming one of the most talented
and passionate writers of our era'
– Steve Silberman, award-winning author of *NeuroTribes*

'Miraculous memoir . . . His portrait of loving parents
raising three neurodivergent children on poetry,
punk and puffins is profoundly moving'
– *Observer*

'Dara captures the essence of why we all
spend time in nature. The excitement, the peace,
the solace, the escape from all our problems'
– Mya-Rose Craig (AKA Birdgirl)

'Like reading William Blake or Ted Hughes, it really is a
strange and magical experience . . . surely one of the most
talked about nature books, or any books, this year'
– *Daily Mail*

'Dara's diary is urgent and compelling: in language as
fresh, precise and vivid as his own observations, his is a
passionate call to arms – urging us to pay attention to the
extraordinary natural world all around us'
– Caroline Lucas MP

'A diary of delight and immediacy . . . Dara McAnulty
reminds us that simple joy can actually help save the world'
– Kathleen Jamie, *New Statesman*

'Dara's writing is evocative and poignant, it stirs something
deep within you and evokes a desire to do all you can for
the natural world which he so beautifully portrays'
– Bella Lack

'An impassioned and original plea for protection for "our
delicate and changing biosphere" . . . lovely and remarkable'
– *Spectator*

'A beautifully written, profoundly important classic of nature writing that will ignite a passion for the wild in every reader. A stunning achievement. I adored it'
– Lauren St John

'The fanfare is wholly justified: this is an astonishingly assured book for one so young'
– *Bookseller*, Book of the Month

'McAnulty's writing glows with his deep sympathy for the natural world'
– Tim Flannery

'Breathtaking'
– Philip Marsden

'Beautifully written diaries . . . Dara can teach us all a thing or two about truly glorious descriptive writing'
– *Irish News*

'His observations are unmatched among his peers . . . listen to what your body and your brain are telling you: thank goodness for Dara'
– *Big Issue*

DARA McANULTY

DIARY OF A YOUNG NATURALIST

2

Witness Books, an imprint of Ebury Publishing,
20 Vauxhall Bridge Road,
London SW1V 2SA

Witness Books is part of the Penguin Random House group of companies whose addresses can be found at global.penguinrandomhouse.com

First published by Little Toller in 2020
This edition published by Witness Books in 2021

www.penguin.co.uk

A CIP catalogue record for this book is available from the British Library

ISBN 9781529109603

Printed and bound in Great Britain by Clays Ltd, Elcograf S.p.A.

The authorised representative in the EEA is Penguin Random House Ireland, Morrison Chambers, 32 Nassau Street, Dublin D02 YH68.

Penguin Random House is committed to a sustainable future for our business, our readers and our planet. This book is made from Forest Stewardship Council® certified paper.

For my family

Contents

Introduction

As I write, snow sits five centimetres deep. Stubborn against the gently rising winter warmth. The windows I look out upon have been constants in a world that is changing at warp speed. Yet despite the shift, despite the flurry, we are still rooted to home, growing and retracting with each day.

This book, my first book, took flight when the planet was, necessarily, bolted, still. We cradled each other's fears, hopes, frustrations. My mum contracted coronavirus while we launched *Diary of a Young Naturalist* into this strange, new world. As a family, we are 'close as otters', but even our tempers have been tested by this intense and difficult time. We have realised that we are not as introverted as we once thought. Yearnings for friends to walk and chat with, share food with, keep spirits balanced with. This knowing has brought both grief and anticipation for more connected days.

Yes, my book found its wings, but there was a period of holding it up to feel the air first. During the months I held it aloft, my arms grew tired from its load. The weight of the responsibility arrived, tied to many wonderful joys, as I removed other books from my shelf to make space for new, heavy-based structures – prizes that awarded my efforts. Letters, emails, Zoom calls came flooding in.

Gratitude pulsed everywhere inside me, but my legs grew weary.

Holding myself up in this storm became almost impossible. I fell through each day. But then I was lifted to the light by many rescuers, who raised me up and showed me compassion. This young body and mind were hurtling, hurting.

We were all feeling it though, right? As we grasped the enormity of the happenings in the news and at home. The intense emptiness where hugs once were. The simple forgetting of your mask as the military planning of *the* shopping trip fell short. The brief period when I attended school, with its one-way system and constant mask wearing, meant that packing one in my pocket became as instinctive as putting my shoes on.

Of course, there was some light. For me, it was the spotlight kind. It showed every pore; nothing was hidden. I was out there for all to see. Feeling my way turned into a kind of knowing my way. I began to stand stronger and allow myself to bask in the light and accept its soft heat, letting the noise die away. New sounds tiptoed in and spun around me, protective skeins of comfort.

As my book took off it found its own way, breaking away from me. In this time, the eyes of many, many more people were opening to the wonder of nature. The light for them, too, was dizzying; the connection had been so broken before. Some had forgotten how to 'be' outside. I told myself that at least the thread would be spun once again, the instinctual ancient ways could return. I still believe they will, and that everyday modernity can make room for all we are: our roots. I have been told (and I shyly tell you) that my book can help us reroute this disconnect, so we can find our way back to the natural world. To the foundation of who we are.

I love this book and I am also laid bare by it. We now have a complicated relationship. When we nestle ourselves down on the grass, feeling every inch of skin touching the earth, it's a self-conscious existence. The writing of this diary, and its publication, has felt much like that to me. I enjoy it when the spotlight is cast on the issues that my book delves into, rather than the author who lifts the pen. But this vessel still carries a weight, and I am learning to stand tall while I hold it, as I continue to grow, into the light. The words of my favourite poet ring true, and guide me every day: 'Walk on air, against your better judgement'. Let's take root, together.

D. McA.

County Down, 18 February 2021

Prologue

This diary chronicles the turning of my world, from spring to winter, at home, in the wild, in my head. It travels from the west of Northern Ireland in County Fermanagh to the east in County Down. It records the uprooting of a home, a change of county and landscape, and at times the de-rooting of my senses and my mind. I'm Dara, a boy, an acorn. Mum used to call me lon dubh (which is Irish for blackbird) when I was baby, and sometimes she still does. I have the heart of a naturalist, the head of a would-be scientist, and bones of someone who is already wearied by the apathy and destruction wielded against the natural world. The outpourings on these pages express my connection to wildlife, try to explain the way I see the world, and describe how we weather the storms as a family.

I started to write in a very plain bungalow surrounded by families who kept their children behind closed doors, and empty-nesters who manicured their gardens and lawns with scissors – yes, I actually witnessed this. This is where sentences first began to form, where wonder grappled with frustration on the page, and where our garden (unlike any other in the cul-de-sac) became a meadow during the spring and summer months, with wildflowers and insects and a sign that read 'Bee and Bee' staked in the long grasses, and where our family spent hours and hours observing the abundance that other gardens lacked, all of us gloriously indifferent to the raised eyebrows of neighbours that appeared from behind curtains from time to time.

We've moved on since then, crossed the country to make another home, and not for the first time. We've lived in many places during my short life, in a kind of nomadic existence. But wherever we settle, our home is crammed with books, skulls, feathers, politics, unbridled debates, tears, laughter and joy. Some people believe that roots grow from bricks and mortar, but ours spread like mycelium networks, connected to a well of life lived together, so that wherever we go we stay rooted.

My parents, both from working-class backgrounds, were the first generation of university-goers and graduates in their families, and they are still fresh with ideals for making the world a better place. This means that we're not rich materially, but as Mum says 'we are rich in many other ways'. Dad is – and always has been – a scientist (marine and now conservation). He's brought alive the secrets and knowledge that wild places hold and explained the mysteries of nature to us all. Mum's career path resembles the way she crosses a stream: never in a straight line. Music journalist, voluntary sector, academic – she still does a little of all these things as well as teaching my nine-year-old sister, Bláthnaid, at home. Bláthnaid's name means 'blossoming one', and at the moment she's a fairy expert who can give you a multitude of insect facts, keeps pet snails and also fixes all the electrical equipment in the house (which Mum boggles over). I also have a brother called Lorcan – 'fierce one' – who is thirteen. Lorcan is a self-taught musician and never fails to rouse in us sheer wonder and confusion all at once. He's also an adrenaline junkie – think running down mountains, jumping off cliffs into the sea, and generally going through life with the energy of a neutron star. Then there's Rosie, a rescue greyhound with severe flatulence and a brindle coat, whom we adopted in 2014. She's our tiger-dog. We call her the living cushion, and she's a wonderful companion and stress reliever. Me, well, I'm the pensive one, always with dirty hands and pockets stuffed with dead things and (sometimes) animal scat.

Before I sat down to write this diary, I had also been writing an online blog. A good few people enjoyed it and said more than once I should write a book. Which is quite amazing really, as a teacher once told my parents 'Your son will never be able to complete a comprehension, never mind string a paragraph together.' Yet here we are. My voice is bubbling up, volcano-like, and all my frustrations and passions may just explode into the world as I write.

Not only is our family bound together by blood, we are all autistic, all except Dad – he's the odd one out, and he's also the one we rely on to deconstruct the mysteries of not just the natural world but the human one too. Together, we make for an eccentric and chaotic bunch. We're pretty formidable, apparently. We're as close as otters, and huddled together, we make our way in the world.

SPRING

In the darkness my dreams are interrupted. I'm somewhere between swimming to the surface and coming up for air when the flute catches my consciousness. The bedroom walls disappear. The space between my bed and the garden narrows, becomes one. I rise without moving, pinned by the heaviness of sleep. The notes keep falling on my chest. Now I can see the blackbird in my mind, its testosterone arrows flying as the territorial sonatas spread across the dawn. Engrossed in this symphony, awake and thinking, the whirring of my brain begins.

Spring varies from space to space, but for me it's the sights and sounds swirling around my everyday, from sky to roots, that hold the most magic. Spring is the frog that crossed our path at the beginning of our time in this house – our first encounter was a splodge of spawn left quickly on the road, its invisible pathway intruded upon by modernity. Upset, we dug out a watery sanctuary with hope: a small bucket of water buried and filled with broken clay pots, pebbles, plants and some sticks for the entrance and exit. We didn't really know if it would work. (Anything deeper would've needed a digger to break through the boulder clay that we're blessed with in our suburban Enniskillen garden.) But there was another meeting, the following year, when our amphibian friend danced a jig on the grass and was joined by another, leaving us a gift of frogspawn in the bucket-refuge. We were exultant, and our whoops of excitement could be heard from the bottom of the hill, drowning out for a moment the sound of cars travelling to Sligo or Dublin, and even rallying against the background noise of the concrete factory nearby.

The ebb and flow of time punctuated by the familiar brings a cycle of wonder and discovery every year, just as if it's the first time. That rippling excitement never fades. The newness is always tender.

Dog violets push through first, just as the sparrows dig the moss from the guttering and the air is as puffed out as the robin's chest. Dandelions and buttercups emerge like sunbeams, signalling to bees that it's safe to come out now, finally. Spring is all about watching each resurgence. Bláthnaid celebrates by counting daisies every day, and when there's enough to make a crown she becomes the 'Queen of Spring' – if there's some left over, she makes a bracelet and matching ring to complete the trinity. At some point, like magic, there are enough daisies for a whole week's worth of trinkets and charms, so she leaves us all daisy gifts around the house.

I've been told more than once that I was an aurora baby, always awake at dawn. I was born in spring, and my first mornings were accompanied by the sonata of the male blackbird, nourishing a growing body and mind. Maybe its song was the first lure to the wild. My calling. I often think of St Kevin, Caoimhín, picture him standing with his outstretched hand, cradling a blackbird nest until the single chick has fledged. Caoimhín of Glendalough was a hermit who sought solace in nature. Gradually, as more people came to see the holy man, seeking out his advice and teaching, a monastic community grew.

I love the stories of Caoimhín, perhaps because

Caoimhín is also the saint's name I took at my confirmation. Although I now feel that this experience was more a 'coming of age', his name is still important to me, even more so now because his story shows that we just can't help intruding on wild places, and altering the balance between people and nature. Perhaps that's how Caoimhín felt, too, as more followers arrived.

The richness of the notes. I can pick them out, even from the most crowded air space. They are the start of it all, the awakening of so much. The song carries me further back: I'm three, and living either inside my head or amongst the creeping, crawling, fluttering, wild things. They all make sense to me, people just don't. I'm waiting for the dawn light to come into my parents' room. Lorcan is nestled between Mum and Dad. I'm listening for the notes, and they come just as the first slice of light reaches the curtain. Golden shadows unveil the shape I've been waiting for: the blackbird harking from the kitchen extension, a glorious messenger on the rooftop of the sleeping and waking.

When the blackbird came, I could breathe a sigh of relief. It meant the day had started like every other. There was a symmetry. Clockwork. And each morning I'd listen and touch the shadows, not wanting to open the curtains and wake everyone up. I never ever wanted to destroy the moment. I couldn't invite the rest of the world in, with its hustle and bustle, its noise, its confusion. So I listened and watched – the tiny movements of beak and body, the straight lines of the telephone wires, the thirty-second interval between verses.

I knew that 'my bird' was the male because once I'd crept downstairs, just once, to look out and up from the patio doors. It was stark and grey but he was there, and so he was always. I counted and remembered each beat, then crept upstairs again to watch the shadowplay on the curtain. The blackbird was the conductor of my day, every day, for what seemed like a long time. Then it stopped and I thought my world would fall apart. I had to find a new way of awakening, and that's when I learnt to read. Books about birds first, and then all wildlife. The books had to have accurate illustrations and lots of information. The books helped bridge my blackbird dream. They connected me to the bird, physically. I learnt that only male blackbirds sing with such intensity, and that birds sing when they have a reason to, like defending territory or attracting a mate. They didn't sing for me, or anyone else. The loss of that song in autumn and winter was traumatic, but reading taught me that the blackbird would come back.

Spring does something to the inside of you. All things levitate. There's no choice but to move up and forwards. There's more light too, more time, more doing. Every past spring merges into a collage and it's so full of matter, all that matters. And that first memorable spring, so etched and vivid: it was the start of a fascination with the world outside of walls and windows. Everything in it pushed with a gentle force, it begged me to listen and to understand. The world became multidimensional, and for the first time I understood it. I began to feel every particle and could grow into it until there was no distinction between me and the space around me. If only it wasn't punctured by aeroplanes, cars, voices, orders, questions, changes of expression, fast chatter that I couldn't keep up with. I closed myself away from this noise and the world of people that made it; I opened up among trees, birds and small secluded spaces that my mum instinctively and regularly found for me in parks, forests, on beaches. It was in these places, apparently, that I would uncoil: face tilted with concentration, wearing a very serious expression, I absorbed the sights, the sounds.

I suddenly fade out and in, realising that it's light outside and the dawn chorus has stopped. The spell is broken. It's time for school. These days, it feels like things are changing. I'm here, on the cusp of my fourteenth year, and the blackbird, that conductor of my day, is just as important as it was when I was three. I still crave symmetry. Clockwork neediness. The only change is another kind of awakening: the need to write about my days, what I see, how I feel. Amongst this onslaught of life, exams, expectations (the highest of which are my own) come these outpourings, and they are becoming a cog in the cycle between waking and sleeping and the turning world.

Wednesday, 21 March

The arrival of March is a time of emerging colour and warmth, but standing in my garden today is like being encased in a snow globe. The icy flakes bite and snatch away the brightness of yesterday. This cold snap brings with it a tough time for our garden birds. They're our extended family, so I rush out to buy more mealworms from the garden centre just down the road, to top up our feeders outside the kitchen window; the feeders are a good twelve feet away to draw a line between neighbourly privacy and invasion. Just a few days ago our blue tits were prospecting the nest boxes and the birdsong in the garden was a concerto of anticipation. And now this. Birds are resilient, but this dip in temperature has us all worried.

It's hard to believe that we felt the whispering of warmer days – only last week – while we were in the branches of an ancient oak at Castle Archdale Country Park, where Dad's office is based. Many people attribute my love of nature to him. He's definitely contributed deeply to my knowledge and appreciation, but I also feel the connection was forged while I was in Mum's womb, the umbilical still nourishing. Nature and nurture – it's got to be a mix of both. It may be innate, something I was born with, but without encouragement from parents and teachers and access to the wilder places, it can't bind to everyday life.

Dara, my name, means 'oak' in Irish, and sitting up in the branches of that majestic tree, feeling the pulse of a life that has been growing in Castle Archdale soil for nearly five hundred years, I was clinging to my childhood by a twig.

I watch a chaffinch in the garden with confetti flecks on his silver crown. He rests on a branch of our cypress tree, an evergreen now turned powdery-white with snow. The chaffinch's peach-blushed chest puffs out as he's joined by a pair of siskins – one citrus-yellow and black, the other delicately flecked with pewter, a daintier yellow. The robin, as ever, is lording it about, pompously strutting to ward off any usurpers. Earlier, there was a four-males-and-one-female tussle of feathers and pecking heads – robins are so aggressive they're said to sever the neck of any opponent, but I wonder if they do that in gardens so full of seeds, nuts and fancy buggy nibbles. Plenty for all.

A song thrush plays hopscotch in the snow, scrabbling around for the seed we've scattered. The bright red of half-eaten apples is spotted: the thrush pecks, releases juice, I smile. The thrush comes at odd points in the seasons, which is the sort of unpredictability that would've caused me frustration and pain in the past. But now I've learnt to rationalise the unreliable thrush, and to appreciate all the encounters without ties or expectation. Well, sort of.

In the evening we celebrate Dad's birthday as though it's a winter Wassail: we all sing and dance and play our tin whistles (badly), shrieking notes and demanding the end of darker days, calling out for light. Mum has baked him a cake – Victoria sponge, his favourite.

Sunday, 25 March

I find the tail end of winter frustrating, and all this waiting to travel through a portal into colour and warmth brings out my worst characteristic: impatience! Today, though, the heat of the air and the hum and buzz all around allay my restlessness. At last, spring seems to be escaping the retreating shadows of winter.

This morning we're all off to one of our favourite places: Big Dog Forest, a Sitka plantation close to the Irish border, about thirty minutes from home, high in the hills, with pockets of willow, alder, larch and bushes of bilberry in midsummer. Its two sandstone mounds – Little Dog and Big Dog – are said to be the result of a spell cast on Bran and Sceolan, the giant hounds of the legendary Fionn Mac Cumhaill, the hunter-warrior and last leader of the mythical Fianna people. While out hunting, so the story goes, Fionn's two dogs picked up the scent of the evil witch Mallacht, and gave chase. The witch fled and changed herself into a deer to stay ahead, but the hounds still snapped closer, so Mallacht cast a powerful spell that turned the dogs, little dog and big dog, into the two hills we see here today.

I love how these names tell stories about the land, and how telling these stories keeps the past alive. Equally, I'm fascinated by the scientific explanations that geologists blast this myth with: the sandstone of the hills is more robust than the surrounding limestone, and as this wore away through glacial erosion, it was the sandstone that remained, towering above the fallen rubble of the Ice Age.

I spy coltsfoot, bursts of sunshine from the disturbed ground. White-tailed bumblebees drink and collect hungrily. Dandelions and their allies in the daisy (or Asteraceae) family are often the first pollinating plants to flower in spring, and are incredibly important for biodiversity. I implore everyone I meet to leave a wild patch in their garden for these plants – it doesn't cost much and anybody can do it. As nature is pushed to the fringes of our built-up world, it's the small pockets of wild resistance that can help.

Sometimes, ideas and words feel trapped in my chest – even if they are heard and read, will anything change?

This thought hurts me, and joins the other thoughts that are always skirmishing in my brain, battling away at the enjoyment of a moment.

The click-clacking of a stonechat brings me back to where I should be, in the forest, and I watch as the bird seems to drop tiny gravel fragments onto the path. I gaze down as the light passes over the path and realise nothing is motionless. Even a stone pathway can move and change with the light and the silhouettes of birds in flight. Each moment is a picture that will never be identically repeated. I watch, captivated, not worried by what onlookers might think, as we usually have this place to ourselves. I can be myself here. I can lie down and stare at the ground, if I choose to. And while I'm staring, inevitably, a creature passes by the tip of my nose: a woodlouse this time, ambling from nowhere to somewhere. I offer it my fingertip and it tickles my skin. I love the feeling of holding a creature in my hand. It's not even the connection I feel, but the curiosity it quenches. As you look closely, the moment sucks you in – again and again it's a perfect moment. All other noise disappears from the space around you. I move to the grass and gently lower my finger to the blades: the woodlouse disappears in the undergrowth.

Bláthnaid and Lorcan rush ahead to the brow of the hill which drops down to Lough Nabrickboy, while Dad, Mum and I amble, chatting about replacing Sitka with native trees in this special place. Last year, at almost this exact time, we reached the top of the hill and saw the magnificent sight of four whooper swans – the only true wild swans. These gentle, melancholic figures bobbed gracefully on the pool, necks held high. They could have been The Children of Lir: Aodh, Fionnuala, Fiachra and Conn, cursed by their cruel stepmother Aoife to spend three hundred years on Lough Derravaragh, three hundred years on the Sea of Moyle and three hundred years on the Isle of Inishglora.

Slowly and quietly we approached the willow-shaded picnic table beside the lake, and they stayed with us as we sat in silent reverence and awe. We felt so privileged. My heart beat faster, my breath felt trapped in my chest. The birds sailed along nonchalantly, until suddenly the honking and trumpeting began. I moved to take a closer look, shielded by the bare branches of a willow tree. I sat as still as the air, watching the widening ripples made by the birds' readying for flight: wings extended, heads dipped, their legs rotating ferociously as they rose up, ungainly webbed paddles giving them forward thrust and lift-off. Away they went, bugling, like a royal convoy. They disappeared to the north-west, perhaps towards Iceland.

To even hope for a repeat of this encounter would be unheard of, and looking down across the lake, I can see there are no whoopers today. It's empty.

I feel a little melancholy as we make our way down to the picnic table. I find a spot and wait for the hen harriers, transfixed until the light fades. When it's time to go my parents exchange knowing looks – and of course they're right, because I'm morose for the rest of the day, and when we get home I slink off into my room, to write, to mope. No whooper swans today. No hen harriers.

Saturday, 31 March

In late-afternoon light, with wind rising from the sea, we sail on the ferry the few miles from Ballycastle on the north-east coast to Rathlin Island. Guillemots and gulls scrabble the air with screeching and cackling. My excitement is intense.

Today is my birthday, and this morning I lay awake in bed for hours before the actual birth time (11.20am) listening to a screeching fox in the distance. All week I've been like this, intensely excited, nervous, for reasons I may never truly understand. Perhaps it's because I love new places and hate new places all at once. The smells, the sounds. Things that nobody else notices. The people, too. And the right and wrong of things. Small things, like how we'd line up for the ferry, or what was expected of me on Rathlin when we arrived. Though I always do the usual mop-up operation in my mind after any journey, look back and usually think how ludicrous it all was, still the anxiety floods in. Mum assures me that our time on Rathlin will be spent either outside or alone with the family. 'It'll all be okay,' she says.

Eider ducks congregate at the harbour on our arrival, and as we head out to the cottage that we'll be staying in for a few days, my usual dislike of new surroundings softens. This place has something special. It feels so calm here. The air is fresh, the landscape extra-worldly in its abundance. Lapwing circle to our right, a buzzard to our left. The windows are rolled down and the sound circulates through our limbs, stiff from the three-hour drive and ferry ride. We relax and radiate as hares abound and geese honk above. The car climbs above sea level towards the west of the island.

When we reach our roosting place, it looks perfect in the distance: traditional stone with no other dwellings around for miles, and on arrival I jump out to walk and explore. I soon discover a lake with tufted duck and greylag geese. As I walk, hares seem to keep popping up everywhere and my eyes struggle to keep up with all the movement, my brain whirring with all the senses.

I can hear the cries of seabirds in the distance. Gannets fly on the horizon, the squeak of kittiwakes becomes louder. I stand and look out to sea and watch the waves gently swirling, and in the dusk sky a skein of white-fronted geese fly in dagger formation. Although we've just arrived and have a few days here, I start wondering how empty I'll feel when it's time to leave. I feel panic.

My childhood, although wonderful, is still confined. I'm not free. Daily life is all busy roads and lots of people. Schedules, expectations, stress. Yes, there is unfettered joy, too, but right now, standing in an extraordinary and beautiful place, so full of life, there is this terrible angst rising in my chest. I walk back to the cottage in a trance, watching shadows moving in golden fields.

After dinner, song bursts from every corner of the sky and we stop to listen in the twilight. Isolating each and every melody, I feel suddenly rooted. Skylark spirals. Blackbird harmonies. Bubbling meadow pipits. The winnowing wings of snipe. And always the sound of seabirds. We are in the other world. No cars. No people. Just wildlife and the magnificence of nature.

It's the best birthday.

A full moon beams from behind clouds as we watch Venus above the distant houses, and I stand there with numb hands and a numb nose but a bursting heart. This is the kind of place I can be happy in. I wrap my coat tightly around my chest, inhaling it all in, not wanting to go to bed, storing the moment up with all the other memories I keep cached. When I'm ambushed by the anxiety army, when it comes stomping back, I'll be ready to fight, armed with the wild cries of Rathlin Island.

Sunday, 1 April

After a night of good food and music and with birdsong still swirling in my head, I wake to promising weather, bursts of blue emerging from the cloud. The morning sea is calm and dazzling. It's Easter Sunday, and we're heading to the RSPB West Light Seabird Centre, home to the largest seabird colony in Northern Ireland – and not too far from the cottage.

I run around with Bláthnaid and Lorcan before breakfast, searching for the chocolate eggs Mum and Dad have hidden in the cracks and crevices of a drystone wall, under rocks and behind clumps of grasses. This is so different to our small suburban garden, where the eggs are found too quickly! We squeal and run, full of unbridled excitement. We don't have to control ourselves here: there isn't anyone around for miles!

Skylarks are our Sunday choir as we walk out west, the landscape our place of worship, as it always is. It's breezy but bright. I spy a pair of greylag geese nibbling grass by the distant edge of the lake, and by the time we reach them I count eight in total, waddling close to us. They show no fear.

When we arrive at the seabird centre we realise we're half an hour early, such was our urgency to get here. We're met by Hazel and Ric – they've been living on the island for a year and are incredibly knowledgeable and passionate about the wildlife, and so very warm and welcoming. I don't say much, but that's not unusual for me. I always smile and nod, except when talking about birds. But even then, although I might look comfortable from the outside, I'm not. I feel squeezed in the middle. I'm trying to process conversations, always looking for nuances, facial expressions, intonation. It often gets far too much, so I just zone out. My heart beats too fast. Sometimes, I walk away from people without realising. It can all be a bit awkward.

Hazel and Ric walk with us towards the stone steps down to the seabird colony. Mum and Dad are still indulging Hazel and Ric in the usual adult pleasantries – all unnecessary conversation if you ask me! I stride out ahead, to start on the ninety-four winding steps that slowly reveal a rugged cliff face alive with kittiwakes and wheeling fulmars, twisting and throwing and dancing in the air. The sight of it makes my insides wiggle. In a sudden fit of excitement I rush down the rest of the steps and across the viewing platform. I can see the guillemot stacks! The cries of excited birds explode inside my chest. Hands trembling, I borrow a tripod from Ric, set up my scope and peer out to sea.

After only moments of scanning, the monochrome suit of a razorbill comes into focus. It bobs in the water and amazingly, despite the churning waves, stays in line with the group. They're such smart-looking birds, even while swaying at sea. I spot a streamlined northern gannet (our largest seabird) in the sky, nonchalantly swerving – it can reach an astonishing 62 mph when diving to feed, but this is a spectacle I've yet to see. They're beautiful birds with stunning eyes, Art Deco lines and a six-foot wingspan. I manage to catch one in my scope, just about. Everywhere, there is the sound of fulmars cackling and creaking like hags hexing the cliffs and all who rest on them. They're quite amusing birds, vomiting a rancid bright-yellow oil to repel nest intruders. I find them strangely delicate and enjoy watching them sailing to land. The whole scene is mesmerising and hypnotic. The screeching soundtrack is perfection. There are no puffins, but I didn't really expect them yet.

It's incredibly warm today, and I feel so content, at peace. Bláthnaid and Lorcan are getting a little restless though – not everyone has the patience for watching birds. I'm given the option to stay, but head off with the rest of the family for some lunch. It's so difficult to leave, but we make a family deal to return before we leave the island.

In the afternoon, we hike to the beautiful Kebble Cliff. Hare prints cast in mud show its light- and deep-footed antics. They're everywhere again. They mystically emerge from tufts of grass and rest for a while, as if sussing us out, then disappear. Buzzards and ravens intermittently delve and wheel throughout our day and a peregrine passes, swiftly downwards, out of sight. We flush out snipe and woodcock as we walk, their frightened flight taking us by surprise and delight. Skylarks and meadow pipits continue to spiral and ascend, their song reaching into every part of my being, lifting, coiling. All that's missing now is the flutter of butterflies, the whizz of dragonflies. The hum of full spring. I stand still and can imagine how it might sound. I vow to come back for real, in May. Such a day.

All tired from walking and exploring, we drive to the pub to have dinner and play pool. I start filing away each moment in my head so that next week or next month, at some unknown point in the future, when I really need to feel happy, I can recall the details. This almost-mermaid-tail-shaped island has me in its siren spell. I'm completely smitten. It's only six miles long and one mile wide but holds so much – and we've only seen a fraction of it.

Mum and I walk the last mile or so from the pub back to the cottage in search of the rare pyramidal bugle, to no avail. Looking at our cottage and how perfect it looks, my heart aches. Tomorrow is our last full day.

Monday, 2 April

A restful night's sleep is not something I'm familiar with. I find it hard to process and phase out so much of our overwhelming world. The colours on Rathlin are mostly natural and muted in this early spring light, tones that are tolerable to me. Bright colours cause a kind of pain, a physical assault of the senses. Noise can be unbearable. Natural sounds are easier to process, and that's all we hear on Rathlin. Here, my body and mind are in a kind of balance. I don't feel like this very often. It means I can reconnect with myself and my family, which is usually difficult because life can get busy and intense. I amble here. I get to watch birds for hours on my own. I'm free to walk where I want. Free to explore. There's no litter either, no unsavoury anything – unless you don't like animal poo! My curiosity draws me to places like this, where I can pick up guillemot and razorbill eggshells (last year's loot stolen by the ravens), mermaid's purses, shells, bones. At home, we have a thing called 'Fermanagh Time', which means life seems slower compared to most other places. But Fermanagh Time has nothing on Rathlin Time, which is even kinder and more free-flowing.

We wake to wind and grey skies but it doesn't stop Lorcan, Bláthnaid and me running out. The wind chops at our exposed faces, our eyes and mouths fill with salt and freshness. Even in shades of grey and black, the sky here holds such light and space and colour. It doesn't have the heaviness of a suburban sky, perhaps because there's just so much expanse. We scope out the lake again, where we spotted the greylags yesterday. We run and run. There are no hares this morning. They're probably undercover, hunkering in the gale. The lake is writhing with wind but empty of birds.

Out of breath and battered, we return to the cottage to be told by Mum that the ferry is cancelled. Joy! I hope the weather never improves and start dreaming of being stranded on Rathlin. During breakfast, I remind everyone of our deal to return to the seabird centre before leaving the island, but instead of walking through sheets of rain we agree to drive the short distance.

There are fewer birds today: a small pod of razorbills undulating on the turbulence, a couple of greater black-backed gulls. Despite the weather, I raise my head to the sky and breathe the tiny details in. A solitary gannet scythes the sky and its cantering cries synchronise with my heartbeat – Orcadians (the people who live on the Orkney islands) call them Solan Goose, or Sun Goose, and as the rain falls I feel the warmth of its lamenting calls. All too soon I feel Mum's hand on my shoulder – I haven't realised how much time has passed.

We head up to the seabird centre for mugs of hot chocolate, my skin flushed and prickling with the heat of the indoors. My thoughts flit in and out of time as Mum and Dad talk to Hazel and Ric. Gradually, my fingers relax and I feel less numb, so I tune in to hear that we're going out into the wind and rain again, apparently to look for seals.

The drive to the harbour takes much longer than we all thought. The rain has slowed by the time we arrive, scratching rather than scraping, but I'm grateful for the waterproofs as we make our way to the small beach in front of McCuaig's Bar. The seals are not difficult to find: there are six in the turning waves. We watch eider ducks too, drifting. The males' striking plumage seems outlandish against the moderately adorned female. Oystercatchers, redshank and a single sanderling peck through the seaweed, while more heads are bobbing further out, long, spindly legs dancing through the washed-up kelp. There's a seal with a strange red protrusion in its body: a wound made by plastic, healed over but with whatever the object is still lodged in place. The sight fills me with a solar flare of anger. How can we treat wildlife like this?

To try and cheer us all up, Mum and Dad take us to a cosy coffee shop where we demolish crêpes, and they remind us that we are soon heading to the McFaul farm, where we've been invited to feed the lambs later that afternoon. As well as being a farmer, Liam McFaul is the RSPB warden on Rathlin Island, and is working hard to return the corncrake, a critically endangered bird everywhere in Ireland. Last year, a male called but wasn't answered. Liam's nettlebeds might help this year, I hope. Talk of the corncrake and sight of the injured seal remind me that even here, in such a wild place, nowhere escapes human intervention. There is loss everywhere. Loss of habitat, loss of species and ways of life. Though it's being reclaimed here and in many places, it's such a complicated matter. I don't feel qualified to understand it or pass judgement. I know it unsettles me, though. The balance is just never quite right.

These thoughts occupy me into the evening, as we feed the lambs at the McFaul farm. It feels good feeding them. We're not farmers but we all love animals, and now Bláthnaid is talking about being a vet one day!

Back at the cottage, we read by candlelight. Dad starts aloud with Dara Ó Conaola's *Night Ructions*, Mum follows with some poetry, until one by one we all drift away to sleep, protected against the crashing waves and din outside.

Wednesday, 4 April

The morning dawns quietly. The wind has gone, which means we're leaving. The business of tidying and packing keeps my mind busy, but a feeling inside is tumbling in jagged circles. We rush for the ferry, late. With a heavy heart we're out to sea. There is no giggling, no pointing out beyond the waves. Subdued silence. In Irish this feeling is called uaigneas. It is a deep, deep feeling, a condition of being lonely.

We had found and lost something, so quickly. Maybe I'm losing part of my childhood, too. There's a Rathlin space inside me, mermaid-shaped, and it needs to be filled again.

Saturday, 7 April

There's an oppression this morning and all day, weighted, surrounding me. Even though there are so many good things happening outside in a garden full of song and activity, my mind has settled between melancholy and heart-racing anxiety. I feel trapped in suburbia. The wind and rush of air, of a wild place, swirls through my daydreams and night stirrings. The anxiety army is marching but my defences have failed me. I scrabble around the fog in my brain, trying desperately to find a memory, an image to ease the tears and confusion and frustration. I pull the quilt over my head to stifle it all, and fall again into another fitful sleep.

Sunday, 8 April

I haul myself back into the world, grudgingly. Even the enticement of Claddagh Nature Reserve doesn't quieten my inner angst. And I feel right to have been brooding because when we arrive, on the forest floor, where the anemones usually flourish, is a huge gouge of dumped mud and stones blotting the ground between the river and a carpet of wild garlic. I rage inside. A tell-tale digger, parked in an empty building nearby, is the last piece in the jigsaw.

We walk on in anger, and although I spy buds on the trees and golden-leaved saxifrage covering the high banks, all of it is cold comfort. The chiffchaffs are warbling too, but I brush their chirpings aside.

As the day grows warmer we decide to head to Gortmaconnell Rock, a wilder place, part of Marble Arch Caves Global Geopark. It's one of those places where no one ever seems to go, at least not while we're there. We claim a few places in Fermanagh as our 'playground', and this is one of them. I spot my first butterfly of the year, a very worn peacock. A flutter clamours around my chest, teasing apart the knot of tension. I can inhale and exhale a little easier. I run from the bottom to the top of the Gortmaconnell summit and feel the wind breaking apart the turmoil. It surges out into the landscape and I lie flat and look at the clouds. I close my eyes, put a hand on my chest and feel a steadier beat. I sleep for a while; everyone leaves me alone. Those fifteen minutes are more restful than all the interrupted hours of sleep I've had this week.

Wednesday, 18 April

The fourth 'report card' of the year has kept my feet from touching soil and grass, and locked me in a cycle of exams where freedom seems nonexistent. The classrooms at school are claustrophobic. Through the stale air I'm bombarded with fidgets, sighs, shifts, rustles as loud as rumbles. The rooms are bright, so bright that the reds and yellows pierce my retinas. Fluorescents drowning natural light. I can't see outside. I feel boxed in, a wild thing caged.

Though I really enjoy Spanish classes, the room is hideous and makes concentrating absolutely impossible. During almost every lesson I have to remove myself and sit outside the classroom. Sitting still, breathing in and out, vanishing in a maelstrom. Thank goodness for the school 'safe space' – it's a room reserved for kids on the spectrum, or others with needs for a quiet space. Some people think I'm isolated in there, but no. I'm safe. My brain can expand and spill out the burdens.

I like school, I really want to learn. But the learning is so flat and uninspiring. The apathy of the surroundings is intolerable. The things we're learning are as captivating as a dripping tap, while outside the world is so much easier to condense, to understand. You can focus in on one thing: a flower, a bird, a sound, an insect. School is the opposite. I can never think straight. My brain becomes engulfed by colour and noise and remembering to be organised. Ticking things off brain-lists. Always trying to hold in the nervous anxiety. To keep myself together.

Friday, 20 April

This morning I was out of school because I'd been asked to talk at a conference of teachers, an 'eco schools' event. I do love doing this type of work. It's part of my mission, for want of a better word. I must shout from the rafters about how we can all be doing more for our living world, our wildlife, how we can make a difference. I often feel like I'm banging my head on a brick wall.

Today, everyone is friendly, encouraging and genuinely excited to be there, to celebrate what many schools are doing with such little funding. But walking through the cultivated gardens, towards the building where the event is taking place, there's a stench of slurry. I'm here talking about biodiversity but there's a lack of it; neither the smell nor tidy gardens sit comfortably.

My heart starts pounding when I'm asked to stand and speak. I realise I can't see the back of the room – making eye contact with an empty wall far away is an important tool for me when I talk in public. Here, though, the podium is too high. I feel small and try to rise up. The room begins to swell. I feel submerged in water. As I read aloud the string holding me up begins to twinge. I'm about to crash. I keep reading. Smile. Stand for photographs. Talk as much as possible among unfamiliar faces. I then realise I'm still wearing my fleece, which explains the sweat dripping down my neck.

I've no idea how long I've been feeling like this. And when I do finally remove the layer, I get cross with myself because I'm wearing my favourite Undertones T-shirt. Why didn't I do this sooner? My inability to organise myself, to do basic things – like take my fleece off in a hot room – really gets on my nerves. I can't plan for it. I just can't seem to manage without someone prompting (usually Mum or Dad). But then the prompting itself is even more annoying.

I travel home with Dad and we play my favourite music. The Clash, The Buzzcocks, and the rest. We talk a little, but all I really want to do is doze, to try and shut the day away. The music flicks the sensory switch, and the sound travels into me and releases the pressure inside.

Music always makes me feel better, and when we arrive home to Mum's questions and smiles, my obligatory run-down of the day is more upbeat. Afterwards, I escape into the garden with my camera. I don't take any pictures. I doze again instead. No wonder I can't sleep at night.

Thursday, 26 April

As I sit finishing homework in my room, I feel a tingle. I pull open the curtain and push aside my doors. I live on the edge, on the edge of the house, away from everyone else in the converted garage. Mum and Dad always worry that I'm not near them at night, but I'm not a baby, and mostly like it. I stand outside and cock my head to the sky and there it is. A screech. A swift! The first of their hundred-day residency. They're here! All the way from Africa. The most exuberant and spirited of our summer visitors, screaming over my house.

One of the most important single moments in a swift's life is finding a nesting place. But people like my neighbours sterilise their gardens and put plastic or metal spears down the middle of their eaves. This attitude prevails everywhere. It's the norm to stop wildlife thriving in the gaps of our homes and office buildings. And the whole poo thing is ridiculous! It's the standard complaint, how dirty birds are, and justifies the loss of habitat, right on our own doorstep.

For now, the lone swift jubilantly churns – a scout, perhaps feeding, not yet paired and searching for a partner, waiting on the screaming parties to tussle and jostle for territory. It's so hard to believe that many of the babies just fly out of the nest and set off on their massive journey alone. Astonishing. I muse on how much we humans depend on each other for survival, and how wild species are at our mercy for survival. I shudder in the evening coolness. The swift has moved on, leaving behind an empty sky as darkness falls.

Before I go to bed, I eye a modest green stalk, shy against the brash dandelions. A tiny pink bud, the first cuckoo flower, lady's smock. Fields were once covered in this delicate, unassuming spring flower, which is still the chosen resting place for the eggs of the orange-tip butterfly. Tiny orange microdots. I'll check all our green stems later in the year, but I haven't found one despite years of looking. Maybe it has something to do with the field and neon-slurry-greens visible from our kitchen window.

Thursday, 10 May

I take my camera into the garden to photograph a dandelion with its blooms inside out, like a windswept umbrella. It caught my eye because I love dandelions. They make me feel like sunshine itself, and you will always see some creature resting on an open bloom, if you have a little patience to wait. This vital life source for all emerging pollinators is a blast of uplifting yellow to brighten even the greyest of days. It stands tall and proud, unlike all the others opening and swaying in the breeze. The odd one out.

The cuckoo flowers are now plentiful, too, and the first common orchid has burst above ground. I wonder if we'll have more than last year: thirteen magnificent orchids. All of a sudden, raindrops fall from the few clouds above and plonk on all the other open-topped dandelions. The only one to escape unscathed is the one that caught my eye.

Dandelions remind me of the way I close myself off from so much of the world, either because it's too painful to see or feel, or because when I am open to people the ridicule comes. The bullying. The foul-mouthed insults directed at the intense joy I feel, directed at my excitement, at my passion. For years I kept it to myself, but now these words are leaking into the world.

I lift my face to the rain and let cloud particles fall on my tongue.

Friday, 11 May

I am buoyed by the life springing out everywhere, in the garden, in the school grounds, even on the streets around the house. My heart crashes less against my chest. I feel in rhythm with nature, and I start becoming immersed in every moment again, letting each wave hit me and seep in.

We decide to take a late-evening walk after Scouts, to a little park in Lisnaskea, a small town not 15 miles outside Enniskillen. It's a balmy evening and the light is hazy with midges, revelling all around, irritatingly to us. Suddenly, punching above the weight of every other song among the reeds and trees, a sedge warbler. It permeates the airspace. I stop to listen. A moment later, a conversation begins between a sedge warbler perched on barbed wire and another on the willow branch. One in shade, the other choosing light, their chirps put into song all the giddy amazement I am feeling. I wonder, sometimes, how other people respond to these encounters. Do they have the same sense of privilege on hearing a bird like the sedge warbler? After one continuous flight from the Sahara, it lands right here to embellish our summer with crackling excitement.

Edward Thomas, who fitted a lifetime of poetry into two years before being killed in the Trenches of the First World War, captures it perfectly:

> Their song that lacks all words, all melody.
> All sweetness almost, was dearer to me
> Than sweetest voice that sings in tune sweet words.
> This was the best of May – the small brown birds.
> Wisely reiterating endlessly
> What no man learnt in or out of school.

Above the bulrushes, a cloud of hoverflies. The light is dappled and sepia. I'm dazzled by the delicacy of the moment. My insides explode, words ricochet outside-in. I hold them close, because capturing this on a page allows me to feel it all over again.

Saturday, 12 May

Today, on a leisurely stroll around one of our local parks – Forthill Park, a remnant of Victorian Enniskillen – I spot something that I haven't noticed before. The rhododendron we love playing in has been dismembered, along with the cold, dark world it once covered. But this spring, amazingly, there are primroses bursting from below the stumps, visible for the first time in – I don't know how many years. Then, in with primroses, I spy a wood anemone, exposed to the air like a forgotten spell.

The heavy rhododendron canopy has smothered the primrose and anemone alike, cast them in dormancy. Yet suddenly, here the anemone is, the blood of Adonis, the blood of the forest that once thrived here before. A relic. A clue to an ancient murder. Evidence of the loss of the forest and bog that covered all of Ireland. The wood anemone grows a mere six-foot spread every one hundred years. I hope this patch is left alone in the light so it can spread once more. A wood anemone in a park, in a town, where children play. An anemone with so much mythology, so many stories, can now reach out again, open minds, touch lives in this century.

Years ago, Mum went to the primary school that borders the park – she went for nature walks, right here. St Theresa's Girls School. Grey pinafore, red rose badge. She's told me how much she loved it – because Róisín, her name, means 'little rose' in Irish. She remembers collecting oak and sycamore leaves, pine cones, conkers. All the children would lay their finds on a nature table – I wonder how many schools have a nature table these days. I know mine doesn't.

The swallows are exultant here, etching the short grass. I lie down and look up at 'The Happy Prince' – it's really called Cole's Monument, dedicated to General G. Lowry Cole, a soldier and politician in the nineteenth century. But in our house it's called The Happy Prince, after the Oscar Wilde story. When he boarded at Portora Royal School, Enniskillen, Wilde must have looked out at the grey tower and dreamed up his precious story of the boy-statue who befriended a lone swallow, left behind in winter. In the story, the prince witnesses the horrors of the world below him, and asks the swallow to pick away at the gold leaf and jewels that adorn the surface of the statue, to tear them away and give them to the poor. When he is stripped of all his beauty, the Happy Prince is taken down and melted in a furnace. All that remain are his broken heart and the dead swallow. These are lifted, swallow and heart, into heaven by angels and declared the most beautiful things in the city.

The story always makes me cry. It makes all of us cry. I sink deeper into the grass and watch the shadows of swallows and listen to their bubblings.

Wilde despised Portora, and I too spent a soul-destroying eighteen months there. I can't imagine why I wanted to go. Perhaps it was because one of Ireland's most famous writers also went there: Samuel Beckett, who loved it apparently, maybe because he loved sports. For me, every day was agony. I hid it well. The bullies were powerful boys, popular, sporty, and lies tripped off their tongues like diamonds. Dark diamonds. Bloody diamonds. I sit up suddenly, heart thrashing and pounding. It's too painful to even think about, still, even over a year later. I'm glad to be gone from there.

I return to the single wood anemone, so alone, but all the more beautiful for it.

Sunday, 13 May

When you visit a familiar place, it's never stagnant. There's always change, and every new day brings a tilt, another view, something that previously escaped you. That something can be as innocuous as a stone wall. Of course, stone walls could hardly be called that – so much life emerges from the cracks and crevices. Wait and watch a stone wall and I promise you, what emerges is a performance, reserved only for those who stop and look. Today, though, it wasn't what was on the wall or inside it, but what was over it.

We'd been walking for quite some time in Killykeeghan Nature Reserve, a small and secret place not far from home.

It's another of the places we always seem to be alone in. Today, we were searching for orchids, listening for the cuckoo call, running over limestone pavements to see if we could spot some mammal scat.

Bláthnaid loves peering over walls, and is drawn to these moments. She has a sixth sense. We both feel it. And she stopped at the most perfect peering spot, because behind the stone wall of an ancient cashel was a hidden pond, reflecting the sky and squiggling with shadows galore, darting in and out of the light. A convulsing mass of tadpoles, and with them the epic cycle of life, anticipation and fascination. We clambered over and surrounded the muddy edge, peering in delight.

The water is bubbling with methane, which makes me think of folklore, of will-o'-the-wisps and banshees, dancing flashes of red light emanating from the decomposition of organic matter. My dad remembers seeing them in Tamnaharry on his great-uncle's farm, dancing in the dark. These days they are rare because drainage and farmland 'improvement' have claimed most of our swamps, bog and marshlands. Whether it's bioluminescence or the combustion of methane, it's wonderful to let the mind wander off with banshees and will-o'-the-wisps – folklore and stories are so often inspired by the strange and the beautiful in the natural world, and all these stories bring nature, deeply, into our imagination. Plus I just love staring into ponds, so it must be good for the mind. My head is pretty hectic most of the time, and watching daphnia, beetles, pond skaters and dragonfly nymphs is a medicine for this overactive brain.

Ripples appear on the surface tension, from no obvious source. I feel the light sprinkling on my head turn into larger raindrops, breaking my trance as they drip from my brow down my face. Bláthnaid and I wander off to find shelter by a hedge, but when the rain stops she goes back to Mum and

Dad while I carry on in a different direction, alone.

As the globe turns, there are things we reach for at certain times. Today, I wanted so much to hear the cuckoo – a need for seasonal 'firsts' is strong in me. The first of everything is very special. In my fervour to hear it today, I realise I've wandered quite far away from everyone and find myself in a secret spinney of hazel and bluebell. You know when you forget a place and remember it all at once? Being in the spinney takes me straight back to being a toddler, trampling the lilac blooms before Mum whisks me up. Then fast forward a couple of years and I'm fidgeting through a cow pat for dung beetles, and climbing mossy banks searching for unknown things. It almost brings tears to me. Being alone, it's peaceful enough to feel the past, and to feel it overlapping with the here and now of musky scent, trickles of light passing through the canopy.

The verdant-lapis light illuminates a path through the bluebells and hazel, a secret way. Sometimes it's good to have a path, for fear of the wrath of faeries, who are said to live inside the bells of these wildflowers – some say that the ominous ring of the bluebell, if heard, will spell death to whoever bares those unfortunate ears.

I tread softly on the wood path. The bulldozing of early childhood is gone. In its place, there is reverence. A bluebell wood takes much longer than our time on earth to get to this carpet of bloom. It is precious and ancient and magical. And it arrives like clockwork, if left alone, casting a charm on so many open hearts. Here since the Ice Age, the bluebell takes five whole years to grow, from seed to bulb. A labour of slow and perfect growth.

A mantle of bluebells, the cycle of spring, and amongst it all, quite suddenly, making me jump out of my skin, the cuckoo sings loud and close. I decide not to chase it, though. I listen. I smile with relief, knowing all is well in this place.

Friday, 18 May

I'm flicking restlessly on our garden swing. It's full sun and the small enclosed space is brimming with birdsong and hiving activity. I hop off and peer into our wee bucket. I remember the day we filled it with stones and old bits of clay pot and impatiently let it fill with rainwater. We added a cupful of murkiness from the pond at Dad's work, some native oxygenators, and the magic brew grew life. Water fleas first. Within a week, snails. Water beetles followed. Then dragonfly nymph and the holy grail: tadpoles. Birds drink and bathe in our magic cauldron, while under the surface metamorphosis is alive and well with five whole tadpoles! Squiggly, squirming teardrops, eating algae from the side of our potion pot. If you brew your own cauldron, magic will surely happen.

A spring evening spent watching life in a bucket on your doorstep is pure enchantment. Yes, it absolutely is!

I go in for dinner but soon enough I'm back out, making a beeline for the bucket. It constantly surprises us with its offerings. I especially love watching how different species interact. The springtail goes on a little jolly reconnaissance using the water's surface tension, a thick skin to the insect, which is so small and blissfully unaware of what is patrolling the underside of the tension: in the gloom, the water boatman is propelled by hunger on its two oar-like legs, swimming upside down, with its piercing mouthparts, or stylet, ready to attack. The action between the water boatman and the springtail is magnificent. The springtail is so fast, driven forward by the abdominal, tail-like appendage called the furcula. The water boatman is so gracefully crazed. It is great to wile away an hour following their antics.

I look again before bedtime, and the springtail is alive and well. But for how long? My mind skips because, well, I'm too old for my body to be seen skipping into the house. I go to bed happy. We're told that childishness is wrong, bad almost. I mourn a world without such feelings. A joyless world, a disconnected one. I push the feelings aside. As I close my eyes, all I can see is scuttling tadpoles, springingtails and a lurking water boatman.

Saturday, 19 May

Before breakfast, I check on the cauldron again. The water boatman is still lingering but the springtail has gone. I don't question it or wonder, it's just gone. I count and breathe with a sigh of relief: the same number of tadpoles swim around the clay-pot fragments, and one is resting on a piece of wood which fits diagonally across the pond. I have to tear myself away as we're making a two-hour journey to Downpatrick to run some errands with Dad, and then on to Inch Abbey together, where one of my favourite corvids is nesting in plain sight.

It feels like a summer evening except for the buzzing and hum, and the distant screech of terns skimming over the Quoile River to the south-west. Butterflies are everywhere. A clattering of jackdaws rises from the ruins of the Cistercian Abbey. This is something special. As they swoop silently around us, we hear other noises, not quite caws. I explore the crevices of the stone walls and find chicks guarded by twigs. Then, the sounds seem to come from everywhere. I step back and watch the parents tirelessly dart in and out of hidden chambers to feed their young.

The black shapes of jackdaws on the feeder at home always take me by surprise. They look so out of place, teetering shakily on the edge. They eat the fat balls delicately, unlike corvid cousins (especially rooks), who snatch and grab and fly away. They're such highly intelligent, sentient birds. They look into the human eye and search it for intention. They can also learn tricks. What amazing creatures, with wonderful shiny, charcoal and night plumage.

In Celtic mythology, there's a story about a flock of jackdaws who pleaded with the king to let them enter the town to escape the bullying rooks and ravens. The king refused, but the jackdaws persisted and found a lost, enchanted ring which had previously kept the province of Munster safe from Fomorian attack. The king changed his mind and the jackdaws were let into the town as avian citizens.

I do love these stories. They enrich my life as a young naturalist. Science, yes, always science. But we need these lost connections, they feed our imagination, bring wild characters to life, and remind us that we're not separate from nature but part of it. Avian citizens! Why not?

Saturday, 26 May

It feels so good to be back on Rathlin Island again, for the late spring bank holiday. We're staying in the same stone cottage and head straight out to the seabird centre for the afternoon. There are a lot more people compared to last time, so before we head down to the viewing platform, Mum takes me aside. We exchange code words and hand squeezes. I build an imaginary suit of armour around myself and move forwards into the throng, senses popping like corn kernels.

When you first encounter the cliffs here during the breeding season, between May and July, everything gloriously slams into you at once. The not-quite pungent smell. The kaleidoscope of sounds. There are thousands of birds: guillemots, kittiwakes, razorbills, fulmars and puffins, all wheeling or diving, patrolling and protecting, sauntering over the shoulder of the stack. Mind-blowing. Magnificent. This is a place vibrating with survival and endurance. I feel tickled and almost hysterical, but must take it all in.

I try to focus on each species, starting with a fulmar, dozing and waiting, a queen on her throne, alone yet protected by the shadow of wings constantly flying past. She's like the Buddha in a trance, conserving energy, settling on the spot. Then the congregation of guillemots catches my eye, one heaving mass – safety in numbers – that completely cover the stack (the birds and the guano). The razorbills are cajoling each other, craning their necks and clacking their beaks, snuggling into gorgeous sleek plumage just as a monochrome mutiny breaks out amongst them, a fight for territory. The kittiwake pair stick together on the cliffside and in the air. These ocean-faring nomads seem like the softest of gulls, but must be so hardy and tough to endure half a year far out at sea – the young birds only return to land when they are two years old or more. And here come the little waddlers, puffins! Slits for eyes make them look like sleepwalkers, hefty on their small bodies as they make their way across the green grass. It all seems like such an effort for them, but they're determined and charismatic – I imagine them alongside the Wizard in Oz as they bumble from burrow to burrow, diminutive inspectors. In flight, amazingly, they can reach 55 mph by manically flapping four hundred times a minute.

My grin is stretching outwards, from me to the cliffs, as if connecting to each and every wing and beak. I even decide to start my new challenge of talking to other people, interacting. Here, surrounded by this, it's easier. I'm in my natural habitat, and sharing it all with others feels so good.

Sunday, 27 May

I wake with a dry mouth and dry eyes from lack of sleep. I need to find the excitement, the rush. I need to find the ability to move through the day, without grumpiness and self-entitlement. Find the joy in the unknown. Because maybe all of life is unknown and we are grappling in the dark, and at least I have the comforts that so many have not. I have family. I have warmth. I have so much love. It will be okay.

What a day it was yesterday – I cantered up the steps of the viewing platform, the heat and the air mixing with the wind and birdsong in my chest. We went to the pub afterwards for dinner, and watched the crimson sky and the sun slink into the sea. We talked and raised our glasses, and then I felt the tide going out and coming in and going out again when Mum and Dad started talking, chose this moment to tell us we'd be moving. Moving house. Moving county, landscape, people. Moving.

Mum said she feels we need to have a new start, a new school for Lorcan and me. Dad wants to be closer to Belfast, for work opportunities, and to be closer to Granny, his mum, who lives on her own now that Grandad has passed. I nodded. I understood what they were saying. I understood but my throat was burning as I breathed in the salty air. I pushed it all as far away as it could go. Denial. Confusion. Mum and Dad's eyes darted between us and each other. Afterwards, they knew to leave me alone. But as we walked, Mum stopped and hugged us all, and without speaking we made our way towards the cottage and the eeriness of the unknown.

The electricity in my head is only just starting to fizzle out. I get out of bed to a day of scorching sun. After breakfast, as we walk towards Rue Point, there is an overwhelming stench of decomposing kelp and two dead goats. It's intense but doesn't dull the glimmering ocean-island expanse.

I rest on a bank full of sea thistle and watch pipits flutter among the rocks, scanning for more life with my binoculars. The sun is so strong I have to squint to make out the shapes up ahead: grey seals sprawled over rocks. Basking, a little scratching, hardly fidgeting. I feel envious. Not only do they lie on rocks from dawn to dusk, but all it takes is one big heave to plunge themselves into the murky depths to feed. From stillness to full motion, no preparation, no in-between. I pick out individuals, compare behaviour. They have such distinct personalities. In a few months they'll begin their breeding activity. For now, they're resting while they can.

In 1914, grey seals became the first animal to be protected by government legislation. But the Grey Seals Protection Act didn't end the conflict with the fishing industry, and killing continued. Thankfully, public outcry in the late 1970s stopped further culls. But as the biologist Lizzie Daly reported in her short film, *Silent Slaughter*, dead seals were found shot on beaches near Scotland's salmon farms in 2018, so it is still a controversial issue.

I feel sick to think of blood running over these rocks. I shake out the thoughts, turn my gaze outwards. I keep a fair distance from the seals and find great satisfaction in watching the soap opera of flippers as they haul up, protecting space, wriggling and nudging. A fantastic silent movie. I relate to their need for personal space, and their antisocial behaviour. The wind changes and the smell overwhelms me. Enough. Even this most enthusiastic naturalist has to move on. I rise and walk towards some other sparkly thing of interest.

The day passes among the cackling ravens and the beginnings of late-spring wildflowers: clover, buttercup and sea campion. I lie on the grass and watch the cumulus, bulbous against the blue. This weekend break in Rathlin is so short, too short. My life seems to come in bursts, close together, crowding in, diminishing freedom. I relish these moments of quiet and solitude.

When dusk comes it is a bruised blackberry sky. The air is cool and fragrant with hay, and before it gets too dark we start driving around the island looking for the spot that Liam McFaul has told us about, all of us listening intently for a sound that was once so common that it could be heard in inner-city Dublin and every field and farm across the British and Irish isles. We stop at the side of the road and wait. There is a stillness in this place, and the silence is deafening. I hear my heartbeat and feel it exploding out through my ears. The anticipation leaves a metallic taste in my mouth. Dad is about to hit the start button on the engine when the craking begins, clear and quaking as a ratchet. A corncrake. It sizzles against the bleating of lambs and moaning of cows, another wild song sacrificed to the agricultural soundscape.

The crops were once cut late, allowing the corncrake pair to breed and raise young. This way of farming has been replaced with more intensive silage-making through spring and summer. This different seasonal rhythm conflicts with the birds – and the unthinkable happens, a life is cut short by the blades. Imagine it. Every egg cracked. The future of the species in this place, in any place, is broken. Gone. A human in the driving seat, of course.

These days, just the male calls out to the infinite skies. He crakes and keens with no mate to return the sound. We sit in silence and listen, and everyone inside the car is smiling.

I love my family, but in that moment their smiles make me want to scream. How can they? I don't share in the joy. A tear moves down my cheek. I creep out of the car, close the door as quietly as I can and walk towards the sound. Such a tiny patch of earth, and yet here it is, padding among dry reeds.

'I'm sorry,' I whisper.

The bird ignores me, it keeps crexing, and it will keep crexing until the season is over. Night after night. Relentless.

I feel such loneliness and despair watching it, listening in. A surge moves through me. I have to do something. I have to speak out. Rise up.

The sky falls dark, I return to the car. The corncrake is still crexing at the night.

Friday, 1 June

The school week is over and I'm still haunted. I sit on the swing watching the adult garden birds to and fro at the feeders, to eat and dig around in the earth before flying off to feed their young.

I have a weight on my tongue, and I've had it most the week. I haven't been able to speak. School is gearing up for exam time, yet again. Apparently these are the 'more important' exams because they'll have an impact on which GCSEs I can choose. The exams are no problem to me; I actually like sitting tests. I like the challenge, sort of, but they just seem to roll around so quickly and we don't learn enough new things in between. It's so frustrating and tiring. If I didn't write, if I didn't have a way to sort through and filter the fluffiness, the haziness, the overwhelming noise that constantly surrounds me, I think I would implode. All the pressures would crush me. Yet, I'm here and it's a Friday evening, and we're going pond dipping tomorrow.

I lean out of the bedroom window and watch intently as the harrying shapes flit at two-minute intervals. Back and forth. Diligent parents. No rest. It's a joyful time. The fledglings will be out soon, and the garden will be all action. A male bullfinch lands on the wall (we had one this morning too). Its rotund, coral breast is flamboyant against the grey of the stone. It's an ungainly form, plopping down to pluck the seeds of the dandelion heads. He repeats this a few more times and is joined by the dusky-pink-chested female. They converse in bubble and squeak. The male's silver back-plumage comes so close, I could reach out and touch it. The tail flicks closer still. I hold my breath, prepare myself, and at that exact moment a lawnmower drones, our encounter sucked away.

Saturday, 2 June

I run through the long meadow, the heady scent catching on my clothes. I stop at the great oak at Castle Archdale and rest my cheek on the bark. I feel the aged, rough skin, the protective layer. I hear it breathing, our rhythms intertwine. I close my eyes.

Three hundred years to grow, three hundred years living in fullness and three hundred years to die. The thought of it makes me feel as small as the ants scuttling up the skin of this mighty creature.

It has been supporting ants and hundreds of other species for nearly five centuries. I sit on the grass with my back resting on the trunk and look up into the canopy. The leaves shimmer in the breeze and my body lights up. A chaffinch's two-note beat starts the rest of its kin singing, and they all play together in the branches. A private performance. I give it a moment and leave before it is disturbed by some unwelcome racket in the distance. I feel smug. I left when the timing was perfect, and skip back to the others at the pond.

The sky looks daunting from here, a flourish of burgeoning clouds. They seem to canon down from the sky without us noticing, rolling in from some part of the blue we didn't see before. The heavens open and close in about two minutes, and then the flickers of light dart before our eyes: dragonflies, their silken wings etched with the maps of the Carboniferous (their wings spanned six feet when their ancestors flew with dinosaurs). They zoom still, like turbo-boosted flecks of light, wings catching the light and showing us glimpses of eons past.

I spy a common hawker engaged in aerial dogfights, striking out after flies, catching them in a spindly cage of legs. Two red damselflies land on a leaf and they contort in courtship to form a heart: the male clasps the female behind her head, depositing. They fly off together, still attached as another tries to interlope.

The rain stays away, so we fill the dipping tray and catch caddis-fly larvae, pond skaters, ramshorn snails, whirligig beetles and a leech. They squiggle, squirm and dart out and away from each other, squeezed together in a pond-dipping pinball tray. Our five pairs of eyes, child and adult alike, shine with delight. In this moment, each of us is connected with the creatures on the little tray, and every other living thing that moves around us in the late-evening sun.

Tuesday, 5 June

The garden has blossomed in the warmth of these late spring days. So much light and sunshine, compensating for the heaving tiredness and exasperation that comes, for me, at the end of the school year. Friendship has always eluded me – what is it anyway? A collection of actions and words between two people or more, people who grow and change anyway. It's a good thing, apparently. That's what some people say. I don't have any experience, though. I mean, I play board games with a group at my school. We play, we deconstruct the game. We don't 'talk'. What is there to say? Sometimes, I feel that if I start, I might not shut up. That has happened, lots of times. It doesn't end well. Kids in my class, they walk around town together, they might play football together or whatever other sport takes their fancy. They don't talk, though. They smirk and snigger at anyone who is different. Unfortunately, for me, I'm different. Different from everyone in my class. Different from most people in my school. But at breaktime today I watched the pied wagtails fly in and out of the nest. How could I feel lonely when there are such things? Wildlife is my refuge. When I'm sitting and watching, grown-ups usually ask if I'm okay. Like it's not okay just to sit and process the world, to figure things out and watch other species go about their day. Wildlife never disappoints like people can. Nature has a purity to me, unaffected. I watch the wagtail fly out and in again, then step a little closer. Peering in, I see that last week's eggs are now chicks. Tiny bright-yellow beaks, mouths opening and closing silently. This is the magic. This bird, which dances and hops at everyone's feet in the playground, unnoticed by most. Its liveliness and clockwork tail, ticking constantly, never touching the ground. It appears again, and the squawking starts in earnest. I giggle inside, in case someone sees. I have to hold so much in, phase so much out. It's exhausting.

At home, I mooch around the garden and notice the first herb robert flowers, pink wild bloom amongst the verdant. I note it down on my list of firsts in the garden and feel good. I hear Dad come back from work, and with him an injured bat. She's the first of the year and we tend to it – females only have one pup a year, such precious cargo. We feed it mealworms and put water in a milk-bottle lid. The bat's mouth is so small I use one of Bláthnaid's paintbrushes to put droplets on its tongue, hoping it will be something like lapping dewdrops from a leaf or puddle. Dehydration is the main killer of an injured bat, so it's important to get it to drink. But as they're getting better they'll chew up a mealworm like a piece of spaghetti.

They're such innocuous and timid creatures, not worthy of the silly hype that surrounds the movies and Hallowe'en. They're insect-controllers: a single pipistrelle eats 3,000 midges a night. Can you imagine the swarms really ruining your camping holiday if we didn't have healthy numbers of bat populations? It's unimaginable.

The bat sleeps in my room. They always do because it's quiet away from the hustle and bustle of the rest of the McAnulty family. I always sleep so soundly when I have a bat staying in my room. I hear it scratching about in the night and am never afraid, I am comforted.

Friday, 8 June

I trudge to school with a leaden heart: the bat didn't make it through the night, and we didn't lose just one bat, we've lost every generation that could have followed. Her injuries, caused by a cat, were too much and she died, Dad thinks, from infection. I feel so heartbroken. I've finished all my exams but that isn't enough to lighten my spirits.

After school, Lorcan and I arrive home to squeals of delight from Mum and Bláthnaid. 'The fledglings are out! The fledglings are out!' Mum roars with all the childish delight that many of the kids I know have lost before they're eight or nine. The excitement is intoxicating, and it spreads into me and I feel a little airy. We watch through the window as a just-emerged coal tit, blue tit and sparrow rest on the branches of the pine trees, open-mouthed, noisy and boisterous and splendiferous.

Watching the discordant gang, I realise that I won't see them when they're fully grown. Not if we move house.

I've been in complete denial about moving house. Tomorrow, though, we're going house-hunting in County Down, in Castlewellan – a small town six miles from our new school in Newcastle (which Mum and Dad say is too expensive for us to live in). I'm not sure if I feel really annoyed about the whole thing, or whether that tickle I sometimes get thinking about it is a sign of the excitement there might be in starting over again. The opportunity to reinvent myself.

Mum notices my mood shifting. I give her my best broad grin and a hug. It's not easy for any of us, but she and Dad will do most of the work – and the worrying.

Every day, ever since I can remember, Mum has sat me down, sat us all down, and explained every situation we've ever had to deal with. Whether it was going to the park, to the cinema, to someone's house, to a café. Every time, all manner of things were delicately instructed. Social cues, meanings of gestures, some handy answers if we didn't know what to say. Pictures, social stories, diagrams, cartoons. Many people accuse me of 'not looking autistic'. I have no idea what that means. I know lots of 'autistics' and we all look different. We're not some recognisable breed. We are human beings. If we're not out of the ordinary, it's because we're fighting to mask our real selves. We're holding back and holding in. It's a lot of effort. What's a lot more effort, though, is the work Mum did and does still, so light-heartedly. She tells us it's because she knows. She knows the confusion. That's why she and Dad will be doing the worrying about moving, and why Mum will be doing all the planning and mind-mapping, and will somehow know how everything fits together. I'm lucky, very lucky.

Saturday, 9 June

The day is glorious. It's summer weather, I have a new Undertones T-shirt (the 'My Perfect Cousin' one) and I feel good wearing it. I don't know why I love T-shirts with some part of me brandished on them. Maybe it's because it will either scare people away or start a conversation without me having to do anything. Well, either way, that hasn't happened yet!

We arrive at the first house for viewing and Mum hates it, I can tell. I don't like it either. Everything about it is squashed, though we can see the Mourne Mountains from upstairs. The second house is much better but needs a lot of work – the views are extraordinary. Neither of them lights a fire in anyone's belly, though, so that's it for today, thankfully. And because it's still morning we're going to explore the Castlewellan Forest Park, a government-owned forest with native woods, conifer plantation and red kites. It even has a lake and a mountain path. Lorcan and Bláthnaid have already been but it's a first for me. It's so beautiful. I feel a swell of anticipation – if we move here we could live beside a forest. We could be near trees! We might not be crammed in by suburbia anymore. I could ride my bike without worrying about cars.

You see, this is a big deal for us kids. We can't access nature the way my parents' generation could. Our exposure to wildlife and wild places has been robbed by modernity and 'progress'. Our pathways for exploration have been severed by development and roads and pollution. Seriously, you take your life into your own hands if you choose to cycle anywhere in Enniskillen. The roads are congested, busy and unfriendly, especially if, like me, you want to stop and stare. We always have to travel to forest parks or nature reserves for our dose, returning to the starkness of concrete and manicured lawns. To think we could live beside a forest!

The thought keeps echoing and I feel euphoric, almost delirious. We all feel it in the glow of the sun with swallows, house martins and swifts above us, dancing everywhere. So many. I've never seen so many all at once. Not all three together. It's heady and intense. We're all springing, bouncing off one another with sideway glances and controlled smiles. Hoping and holding it all in.

We find a peace maze in the park, created after the Good Friday Agreement in 1998. It has 6,000 yew trees and was planted by 5,000 school children and others from the nearby community. We rage through it until we come to a rope bridge. I stop and get out my binoculars: red kites, three of them, wheeling and soaring, ascending, dropping right over our heads. It's staggering. We gawp at the sky and you can feel our family agreement travelling through us, silently: this might be a good place to live.

Exhausted after the long drive and the day's events, we head back to Granny's house in Warrenpoint, where we're staying tonight. My Granny Elsie has amazing views from her back garden. We can see Carlingford Lough and the Mournes and the Cooley Mountains. Every day looks different there, with subtle changes of colour or the way the clouds sit then disperse on the mountains. Today, the sparrows are chattering and the sun is still high. We decide we need another walk along the beach before we get dinner.

We do a beach clean as we go, but not too much today, which gives us plenty of time for exploring. Lorcan has the best find of the day: a cuttlefish bone smoothed by the sea, silk-soft. The bones, which are not really bones at all but a shell, are usually from the females who die a few weeks after breeding, and the dead cephalopods' skeletons are later washed up on the beach. Lorcan's find has the kind of piddock holes that we normally see in soft rocks and clays, and there still seems to be life inside them so we carry it back to the sea before it dries out. We find another, bone-dry, which we bring back to Granny Elsie's.

Later that night, in the darkness, sharing a room with Lorcan, we talk about the move in hushed tones and excitement, until we both sink like stones into sleep.

Saturday, 16 June

These glorious days have been rolling into one another until they are indistinguishable. The heat is unbearable at times, and the garden is already suffering. The grass is parched and I'm in my last week of school before the summer holidays.

I've just found out that I've been invited to Scotland next week. Mum got a text from Dr Eimear Rooney, who works for the Northern Ireland Raptor Study Group. I've met Eimear a couple of times before, once during Hen Harrier Day and again after a fundraising walk, so she knows how passionate I am, especially about birds of prey. And now she's invited Mum and me on a mission with another hero of mine, the amazing Dave (he does have a surname, but his work is sensitive so I don't want to give it away) who will be leading a satellite-tagging trip for goshawks.

Goshawks! After hearing the news, I sit myself down and flick through the pages of a raptor field guide, stopping on *Accipiter gentilis*, the gentle hawk. I've heard them before, cackling from the depth of conifers in Big Dog Forest, shattering the silence, but I've never actually seen one before. Soon I might be able to hold one! I can scarcely imagine it, never mind believe it. I'm going to learn so much.

I reel myself back and remember that we're making a pilgrimage to Big Dog Forest today. It will be our last visit before we move in four weeks' time. It's all happened so fast but we've found a house. It seems nice. It has rowan trees in the garden, and it's right across the road from the forest park that we explored last week. Although we're all full of exhilaration this morning, I can see the drawn expression of tiredness on Mum and Dad's faces. Mum has been sorting out our school, our educational statements, GCSE choices, furniture removals, and all the while she's still been teaching Bláthnaid at home.

We head out happy but aren't prepared for the emotional connection at Big Dog. It's here I saw my first hen harrier rise out of the trees, and where we heard the goshawk's call. It's here that we've had picnics, conversations, accidents, mishaps. This place has formed me. Soon we won't have many opportunities to come back and meander like we do, wandering lazily through the forest for hours. The new forest park might one day cradle memories too, but I feel treacherous for even thinking it.

Dad, Lorcan and Bláthnaid climb Little Dog while Mum and I sit beside the lake in our spot to watch for the hen harriers. I get distracted by two red admiral butterflies, circling each other in a ray of sun, chasing light, dazzling us. I intuitively look up as a large bird – what can only be a raptor – flies over our heads and into the Sitka trees. It couldn't be. It's impossible! Was it? I manage to shakily zoom in on the black-and-white wingspan, like a pair of barn doors flapping in the wind. Mum and I shriek in disbelief. An osprey! Mum quickly texts a picture to Eimear, and she confirms what we already knew: this bird is too late to be a passage migrant, and its mysterious appearance might be a sign. Could ospreys breed in Ireland once more? We jump in the air but quickly gather ourselves as we wait a little longer for the harriers. As minutes pass I start grappling. It's not just that the hen harriers don't show, it's more than that. We won't be back here anymore, monitoring this spot for hen harriers. I feel grief. A deep grief.

SUMMER

I'm lying on the ground looking up at the branches of an oak tree. Dappled light is shining through the canopy, the leaves whisper ancient incantations. This tree, in its living stage, rooted in sights and sounds that I'll never know, has witnessed extinctions and wars, loves and losses. I wish we could translate the language of trees – hear their voices, know their stories. They host such an astonishing amount of life – there are thousands of species harbouring in and on and under this mighty giant. And I believe trees are like us, or they inspire the better parts of human nature. If only we could be connected in the way this oak tree is connected with its ecosystem.

I often imagine a canopy of leaves above my head, protecting me from the world. More often than not, though, it doesn't work. The humiliation builds into despair. I get completely exhausted by the amount of energy spent taking deep breaths, ignoring remarks, weathering punches. By solstice in June I end up feeling like Scarecrow on the way to Oz, his straw-insides hollowed out. The feeling of complete emptiness is eclipsed by confusion: how can people be so cruel? People my age. My generation. How can they hit, punch, hurl abuse? Who teaches kids to be cruel? Why mock and taunt? Where does all this hate come from?

The pain has dulled though. They cannot hurt me. They don't have power over me, not anymore. I see only beauty in the world, at least I very much try to. The life that surrounds us is so bewitching, so fascinating. Autism makes me feel everything more intensely: I don't have a joy filter. When you are different, when you are joyful and exuberant, when you are riding the crest of the wave of the everyday, a lot of people just don't like it. They don't like me. But I don't want to tone down my excitement. Why should I?

Everything is burgeoning below the oak tree, and Castle Archdale Forest is full of life as I try to fight the emptiness. I'm looking forward to the end of June when school ends and I'm safe again, at home with my family. Grades are always near perfect but that's the easy part; while everyone swaps numbers and arranges to meet during the holiday, I'm the one standing there gaping, bewildered, awkward. I want to belong, yet I hate the notion of belonging. Instead, I spend the summer at home, with every good day spent outside. There are always projects to do, on pollination, the Middle Ages, Beowulf, poetry, music And we love it, especially the road trips. The travelling. The movement. We're never stagnant, unlike at school.

We weren't always so mobile. When I was younger it was just easier if there was nobody else around. I used to have major tantrums, which peaked when I was about seven, and if we spent any time with other families – other parents, other kids – it was hell.

The light is dazzling the ground below the oak tree. I watch it sparkle on the grass as a memory surfaces, emerging in the warmth. It must have been ten years ago now, in Belfast. It was a warm summer day like this one, and we've just left the library on Ormeau Road with some friends. I see a jackdaw feather on the ground, so I pick it up and give it to a girl standing next to me, 'my friend'. I had frequently confused her by my actions, and this day was no exception: she looks at the feather with disgust, then her mum grabs it and throws it away. 'Horrible', she says. 'Dirty'.

I can still feel the heat rising inside me, like particle soup, exploding, crashing. I couldn't control the roar. I roared so loudly and for so long that my brother Lorcan started to cry. Mum, I know, could see the hurt and confusion in my eyes. But what could she do?

I still wonder what that moment must have been like for her, a mother and a friend and a person on a street in Belfast. I remember how it felt when she picked me up, so gently, without blame.

It wasn't the first wild offering I'd tried to give someone but it was the last. Unless they were family, I decided that nobody deserved anything as beautiful as a feather. People just seemed to enjoy nature from a distance; cherry blossom or autumn leaves were beautiful on trees, where they belonged, but not so great when they fell all damp and leathery to the ground, onto lawns or school playgrounds. Snails were an abomination. Foxes were vermin, badgers dangerous. All these strange ideas spun round me like a spider's web, until I was entombed. I was the pesky fly and they were in control. Controlling wildlife, controlling me. But there is joy in the things you love, a power that I'm starting to fight back with, to take back control, fiercely and righteously. Lying below the oak, I can feel it surging below the ground, the roots curling around me, a restless energy feeding me strength.

Thursday, 21 June

The summer solstice starts at 3am. The night is heavy, the air clear and silent as we pack the car and drive towards the ferry port in Belfast. Teenagers are still revelling in their exam results, staggering, helping each other home in the darkness. Mum and I are travelling with Dr Eimear Rooney and Dr Kendrew Colhoun, two of the ornithological experts joining us on our trip to Callender in Scotland. An expedition. An adventure. Proper fieldwork, with goshawks! In the car, I have to suppress a giggle because it all feels a bit like Michael Rosen's *We're Going On A Bear Hunt*.

We arrive at the ferry port in good time, with no hassles. When we travel by plane there is always hassle: delayed flights, squashed seats. Such close proximity to people is all very annoying to me. This is different. I settle down in one of the comfy reclining chairs for a sleep as the grown-ups head off for coffee. I know Mum won't rest, and when I stir, there she is, reading, while Eimear and Kendrew snooze alongside. She smiles at me. 'I'm enjoying the quiet,' she says. 'I never know quiet like this.'

I doze again and when Mum wakes me we're close to shore. We go to one of the viewing points to look at the gulls and see if we can spot anything else. The clouds are lifting, blue is pushing through. I feel so good, full of anticipation. But I can feel my excitement tipping into panic. I wonder what the day will bring. Will I make a fool of myself? Will I be useful? I hope I don't drone on, mechanically reciting goshawk facts. And what if I'm not physically able enough for any of the work? Mum senses my quickening heart. She leans her shoulder towards mine and says she's worried too, but that it will all be okay, 'We're among kindreds.' Bird lovers. Compassionate people. She's right, it will be just grand.

The drive is spectacular and strange: majestic seascapes on one side, bland and luminous fields on the other, one after the other bereft of life, just monoculture. It's more industrial than it is at home, and the sight of it makes me mournful. I wonder what life those green fields deny.

The adults all sound very chipper, chatting away, but I'm more pensive, still contemplating what will lie ahead. I try to envision all the possibilities: the habitat, the craft of looking for goshawks, how to walk or bog-hop the forest floors. We're with experts (and meeting more) who know everything, but it doesn't stop me joining imaginary dots and thinking methodically about how I should talk to people. I rehearse the things I'll say, how to be polite, how to look engaged. My head whizzes – it's hard work picking through the details of the day before it happens. But I desperately want to make a good impression.

My early fascination for raptors has grown into a passion to help protect them. A few months ago, Mum and I bog-trotted and hiked our way through the Cuilcagh Mountains, thirty miles of spectacular landscape, to raise money for a satellite-tagging programme, the first of its kind in Northern Ireland. The work is delicate, secretive, and involves raptors being tracked and monitored so ecologists can learn about how the birds travel, where they nest, their flight patterns and behaviour. Our trip to Scotland is all about training for this, learning from the scientists in Callender. It is also about seeing conservation in action, taking part.

Hunted to the brink of extinction by gamekeepers and

egg collectors in the 1800s, the few hundred goshawks now breeding in the UK are descendants of bred falconry birds released into the wild. I imagine what they'll be like up close, what they'll smell like, what they will feel like. I can't stop thinking about them. Goshawks and ospreys are still being mercilessly killed by people. Shot. Poisoned. Trapped. That a human being thinks and feels it's okay to persecute such beautiful creatures is implausible to me. I feel enraged.

As we drive, I watch gannets dive into the sea and a solitary buzzard hunched on a fence post. Swallows are swooping, and I rejoice in them, as I do every year. Although the car windows are closed, a warmth sweeps through me because I can still hear their bubbling song in my head.

The clouds have all but disappeared. Wispy cirrus decorates the light blue. At the half-way point we stop for coffee – I have a mocha, and when I look down at the empty cup I realise it was a bad idea, it has nearly blown my head off. I gulp down water from my bottle to compensate. Mum drinks two coffees – years of night-owl studying have left her immune to caffeine. We joke that she'd change from angel to demon without her morning coffee, though I reckon she'd be just the same.

By the time we arrive Dave's house at around eleven, I'm all at once full of nerves. I know we'll get on because we both love birds of prey, but I've only seen him on television, satellite-tagging golden eagles, and always feel like this meeting new people. It doesn't help that we've been travelling for six hours and are running late. I breathe deeply. Mum pulls me back for a few seconds, squeezes my hand before I go on.

Dave is larger than life, and his family are there, along with his teammate Simon, who has smiling eyes and a very sharp wit. Dave talks about what we're going to do, and why we're doing it. It's such brave, important work, and

I feel incredibly privileged to be joining them. When Dave hands me a satellite-tagging device, I'm amazed at how light it is, and that such a small piece of technology can monitor the bird's movements anywhere with a network connection, until the solar batteries break down a few years from now. Ideally, none of this would be necessary. The technology, the conservation teams. The constant vigilance. The responsibility. The heartbreak. But so long as raptors like goshawks, golden eagles, hen harriers, buzzards and red kites are persecuted, this sort of human intervention is necessary. Satellite tags not only help build a picture of where the birds travel, but also where they disappear.

We continue on our way in Dave's truck along with his dogs, Simon and another member of the conservation team whom we pick up before reaching the Sitka spruce plantation. It's still late morning when we arrive. We can't drive any deeper into the plantation, so we get out and trek with the equipment. The sun warms my skin. My ears pick out a robin, then chaffinches.

It doesn't take us long to find the first nest: guano marks the ground below it and white feathers are stuck to fallen branches. Reverent whispers travel between us as Simon and Dave delicately lay the tools out. A harness is clipped on, arms and legs go rocketing expertly up the tree at an astounding speed. I stand below and can make out the mewing sound of the chicks above. There, in the distance, the mother starts calling. Her sounds aren't repetitive for now, nor is she swooping at us. The signs are all good, but I hope she doesn't get distressed.

I stare up to the nest, transfixed, stroking one of Dave's dogs to calm my nerves. I can see a bundle being carefully lowered in an orange bag, full of promise. I inhale every scent and every sound in the forest. Pine-dry earth. The creaking branches. There are crossbills somewhere, chattering.

Although I've never seen crossbills either, I hold in my excitement because the goshawk has arrived at ground level. I feel something turning inside me. We catch hold of the rope at the bottom, unclip the bag from the harness and lay the bundle on to the ground. The chick inside looks like an autumn forest rolled in the first snows of winter. The plumage is still downy, feathery constellations shine out all over it. Breathtaking. We are all in awe. It stares at me, deeply. The searching blue eyes and powerful beak are offset, almost amusingly, by its tufty crown of brown studded with stars.

Dave gives me the job of writing in the log book. It feels good being useful, and I take great care to get all the information correct as the chick is tenderly weighed, measured, ringed, tagged. It's a technological ballet, not surgical or invasive, and afterwards the chick sits on the ground as if it were still in its nest, quite unfazed, its head bobbing. Then the process starts over again: two more chicks are lowered in the orange bag. Weighed. Measured. Ringed. Tagged. I find the whole operation mesmerising, this delicate interaction between birds and people. The closeness of one species to another doesn't seem quite right somehow, and yet completely enchanting. Perhaps I'm just not used to it.

Without realising, I start talking to the people around me – Simon, Dave, Eimear and Kendrew. I feel at ease. This is so rare. They aren't teasing or confusing me. I ask questions which are given detailed, intelligent answers, and it feels as if I've been dipped in a golden light. This is what I want to do. This is what I want to be, surrounded by kindred spirits, doing useful things with care, knowledge and clarity. Surely this is enough to quench my overactive brain. Surely this would mean I'd be happy. My endless need for facts and hunger for information don't necessarily fill me with ease. This is different, though. For here and now I am working and seeing and feeling, and it is more than enough.

When we're finished at the first nest, we are led to another site, first through a field of willowherb buzzing and humming with life. I get momentarily distracted by red admiral butterflies and carder bees. I inhale the delicious scent of late afternoon. Moving onwards, we enter another dense plantation, where the terrain is tougher, with higher and thinner trees. It becomes clear that climbing up to the second nest is going to be much trickier. Dave suggests we spread out around the base of the trunk, in case the birds 'jump'. I look up: the trees are like spindly witch's fingers, swaying as if casting spells. Suddenly, four birds jump and one of them is heading right at me. My heart jumps. We scatter as they land, and I step back as Eimear and Kendrew catch them and bring them safely to the ringing station so it can begin again. Wing-measuring with a ruler. Weighing a bird in a bag attached to a scale. Leg banding – a colour band on one leg and a British Trust for Ornithology ring on the other. Attaching the satellite tag with ribbon, delicately on the back of the goshawk. It might sound robotic and innocuous, but to me it's miraculous and exciting.

I start to shiver while I'm watching the chicks. I realise I've not eaten since morning, and we hadn't had the time or foresight to remember a packed lunch. Without food to burn, my body feels the cold seeping in. I keep watching, listening, recording. It helps keep the hunger at bay. Dave asks me to hold one of the birds, and as I bring it close to my chest its body heat illuminates me. I start to fill with something visceral. This is who I am. This is who we all could be. I am not like these birds but neither am I separate from them. Perhaps it's a feeling of love, or a longing. I don't know for certain. It is a rare feeling, a sensation that most of my life (full of school and homework) doesn't have the space for. The goshawk wriggles. I settle it down and stare into its eyes again – as it grows older the baby-blue will change, become a bright and deep amber. I start to imagine it as an adult sailing through the trees, cutting through the air, wings tucked tight, swerving at breakneck speed, building a nest for its young. Will I come back to see it again? I hope this chick will survive.

Once the birds are hoisted back into the tree and carefully returned to the nest, we trek back to the truck and drive out of the plantation. On the way to the hotel, we stop for supper: half-deranged and delirious from lack of food, we are all so red-cheeked and merry that the people at other tables in the restaurant probably think our group (all but me, of course) are drunk.

It's the first time in a long time, probably ever, that my head doesn't stay awake to dissect the day. I just hit the pillow and sleep. Deep.

Friday, 22 June

I wake in the tiny hotel room, light slicing through the thin curtains. There are rooks on the roof clattering above me, and the screeching song of swifts. A good soundtrack to wake up to in a strange place. I feel refreshed and ready for the excitement of what's to come. More goshawk tagging.

Mum and I are staying in a different hotel from the others, so after a shower and breakfast we meet up with Kendrew and Eimear to stock up with lunch and snacks (there's no way we're making the same mistake twice) and then to drive to Dave's house. After a quick chat and a raucous play in the garden with Dave's dog, we set off for another adventure.

The day is so much hotter. Dragonflies are whizzing, grasshoppers are trembling in the grass, and swallows are everywhere. We're in a field bordering another plantation, where the trees meet farmland, and Dave is unpacking a mysterious black box – we're all intrigued. It's a drone, he explains, which if used properly can become an amazing investigatory tool. He sets the machine whirring, it ascends quietly, surges towards a stand of trees, flying nimbly then hovering absolutely still: somewhere among distant branches, a female osprey is sitting on her eggs. Mum and I are captivated as much by the technology as the images of the bird that begin flickering on a screen in front of us, statuesque, piercing us with her eyes. I wonder what she makes of the drone, so quiet and seemingly unobtrusive, only staying above the nest for a moment. The osprey lifts, rearranging her body, revealing a clutch of three eggs. Just like that, a revelation under feathers.

The drone has done its job, in and out in five minutes, astoundingly effective. All too soon for me, the equipment is packed away and the osprey is left in the leaves. It's time for us to walk through the wall of Sitka spruce in search of more goshawks. Dave warns us that the forest is a quagmire, and suggests that we kit up with our waterproofs and wellies. I feel the rush of adrenaline pulsing in my legs as we navigate the terrain, bog-trot over pools, branches and the bright-green sphagnum. Dave's height means he takes one step to our three, and in an effort to keep up, Mum hurries forwards, ahead of the rest of us. I worry, because even though she can stroll up mountains with ease, this is very different. Treacherous really. These plantations, built on bog which ripples beneath your feet – only a frequenter can understand its nooks and crannies.

I watch Dave stride over a particularly large bog pool, Mum behind, readying herself to leap. I see the gap and know her legs won't make it as she steps in and disappears, hip first. Squelch. I feel embarrassed for her and worried, too. Amazingly, she heaves herself out with one leg on the bank, refusing Dave's hand. I know she's probably mortified as she emerges, welly still attached, covered in moss and debris, but she just smiles, empties the boot of bog water and keeps going.

Arriving at a clearing, we can see a female goshawk in the middle distance, circling, calling out. I feel uncomfortable and start worrying that our presence is upsetting her. She lands back at her nest, but rises again and keeps circling, calling out. Dave and Simon conclude that it's best we retreat, respectfully. I take a moment to mentally photograph everything before we go, knowing it's probably the last goshawk we'll see today – and perhaps the last one we'll see on this trip. I drink it all in. The logs on the ground where we sat. The weird patch of nasturtium-bright orange against the lush mossy branches. The way the light is pulsing through the trees. Even that faint smell of slurry from a nearby field.

On the way back we check in on another nest site but find it empty, either abandoned, fledged, or worse. We stand waiting and watching for a while longer, but still nothing. We give up and walk out into the field to eat our lunch in the heat of the day. We settle ourselves on the grass and Dave suggests we go higher into the mountains to see some much rarer birds. 'Would you like that?' he asks.

I cannot contain my eagerness, but it means saying goodbye to Simon, who drives off in his own truck. I shake his hand before he goes and feel such gratitude – I've learnt so much from him in the field, and all this doing is so different from being in a classroom.

As we drive off with Dave, I stare out at the splendid Trossachs, so rugged and wooded – the sight of them reminds me of the Mourne Mountains back home, and my thoughts quickly drift to our plans to move house. Anxiety starts its growl. Unusually, though, I manage to blank it all out. I concentrate hard instead on the beautiful valleys, on the hills rising, the bordering forests filled with pools and streams. I wish my life was full of days like this one. Perhaps it can be.

There are several gates to open and close as we pass through farmland, twisting and turning onto higher ground, before we arrive at another secret location. It's early evening and we're greeted by a swarm of midges as we get out of the truck. I spy what might be some heath orchids, then some common spotted ones, all flourishing alongside knapweed and covered with hoverflies and bees. The sound of water rushes everywhere. The valley sings, heaves and rests. All the expanse, after we've been closed in by plantation forest, is like drawing in a tremendous breath before diving off a waterfall. Freefalling headiness.

It's a relief to follow a path – solid ground. We set up a scope and point up to a steep hill, where there is a vein of solid rock that leads to an alcove. Our excitement is crackling. Dave takes the drone out of his backpack and sets it off again, towards the alcove. We watch, expectantly. The video monitor moves towards the eyrie as the drone sweeps above the rocky face, then hovers. There it is! The camera beams back a golden eagle chick in the nest. What a sight! My laugh ricochets, and then we're all smiling and watching, enchanted. At this age, the parents will feed it every few days, so the chances of seeing them were slim. But there it is: the next generation sitting on the precipice of life and death. We sit below breathing in the moment and feeling the enormity of it. I watch the sun dip behind the valley, happiness flutters in my chest.

Saturday, 23 June

It was Granda who told me about the sounds of the scréachóg. He used to hear them in the countryside as a young man, especially on the way back from the pub at night. These days, the screeching of barn owls is rare in Northern Ireland, as it is elsewhere in these Isles, which means I won't experience the sounds that Granda heard as a young lad. Modern farming and housing developments have depleted roosting owl habitats, while the use of rodenticides has poisoned populations – because barn owls feed on rats, mice and voles. Unless rodenticide is banned completely, the future for them will be bleak.

On our last day in the field, when we spied the female barn owl through binoculars, alone and very underweight, we knew it was possible, in desperate hunger, that she might have eaten her own chicks and was still struggling for food. She was ringed, so Dave and the team will continue to monitor her – we're all hopeful she'll breed successfully next year.

It was a sad, unsettling way to end these enchanting few days of our field trip. But this is the reality. Many birds don't make it. I'm so in awe of Dave and all those that do this important work. They are my heroes, and I'm so lucky to have had a tiny glimpse of what they do. The monitoring is the exciting part, but there is also the burden of waiting for birds to nest and breed, then the fallout and grief when the worst happens. This work must be like riding a pendulum, moving quickly between joy, adrenalin, anguish, anger.

I count buzzards and watch plummeting gannets as we drive to the ferry port. When I sleep against the car window, I dream of blue goshawk eyes, bright-yellow talons and the feel of downy feathers. I hold each memory close. These are the things that will lighten the bad days to come. In three weeks, we'll be moving house – I must hang on to these moments, keep them locked away but always living.

Wednesday, 27 June

The dry spell has continued, the temperatures are still rising. I try to remember when it last rained – was it last month? One hot day melts into another. Apparently it's been the hottest June since 1940. The last days of school are dragging. Other people seem to enjoy free time but it's hell to me. I enjoy daydreaming and thinking, letting my mind wander so it can process things that need to be filed away or understood more. This is part of how I function. But the chatting and the banter that seem to go hand in hand, unless it's about something I'm interested in, make me feel anxious. I just don't know how to play my part. Schools can be extremely bad places to learn if you're autistic. Filtering out noise can be impossible. Focusing and concentrating require so much energy. I'm exhausted by 3pm. Yet, I must come home and do homework then set my alarm, and do it all again the next day. I have to work so much harder than most other 'typical' students. But it has to be done because I want to be a scientist. I want to go to university. The hoops must be jumped through. Apparently, it makes us stronger. Better citizens. I'm not so sure. I think of all the technical advances humankind has made over the last hundred years, yet the way we're educated has stayed more or less the same. With rows of bodies sat rigidly behind desks. Sitting still. Putting up our hand to talk, unless it's a teacher-directed debate (quite rare in my experience). Yet, we accept it. Why? Conformity. Obedience. Duty. And now that our house is starting to fill up with boxes, the uneasiness that is usually left behind the moment I step outside the school gates continues when I open my front door. The mess. It's chaos in here.

I escape to watch birds in the garden: there are fledglings everywhere, alongside the exhausted, bedraggled adults. A rook hops along the hot roof, then cleans its silver beak on the apex slates. One hop, two hop, three, stop. Clean beak, repeat. One hop, two hop, three. In the distance, the wood pigeon is calling out again. It sounds like the song in my head today: 'I don't want to move, I don't want to move.' I can't stop hearing the words going over and over on repeat. 'I don't want to move.'

I flick the imaginary switch in my head so the wood pigeon sounds like a wood pigeon, not me. And I keep those thoughts away by standing up and walking around, pacing, then walking on to watch the tadpoles, which are now froglets. They're basking on the brick and twig bridges we made for them (so they and other creatures can move in and out at will). I hope they leave the rocky cauldron as frogs before we move away to County Down. I lean closer, cast my shadow across the water by accident; the froglets disappear in a blink.

It's unbearably hot, so I take my book to the swing and pull the cover across my face to hide the sun. It's still too hot. I get up, walk around again, sit again. So restless. Mum's watering the raspberries at the side of the house and loudly declares that they're ready to eat. What a relief, something to do! We all plunder in (Lorcan gets there first) and leave with stained hands and lips, my restlessness broken for a few minutes.

When Lorcan and Bláthnaid drift back inside the house, I end up back on the swing pushing myself gently. I start to wonder why life throws me such curve balls, such as this moving-house-shaped curve ball. Is it to help me grow into a 'normal' person? Perhaps, if life shakes things up enough, I'll get used to the mishaps and not worry about them so much. Really, deep down, I know this is never going to happen. I might deal with things in a more visibly 'able' way, but the inner torment will be the same.

After dinner, in the cool of the evening, we decide to head out as a family for an evening walk. Dad drives us to Bellanaleck, a small village about five miles outside Enniskillen, on our side of town. There are still traces of heat in the late evening sun as it sets behind the trees. I watch swallows skim the lake, picking midges off as they go.

I will miss the lakes here in Fermanagh – there is water in every direction. You can't travel anywhere without having water beside you. Lorcan, Bláthnaid and Dad walk on, leaving Mum and me sitting quietly on the jetty, dangling our legs as we watch the swallows part the water with open beaks. After a while, she gets up and heads off to see if she can find the others. I stay, lying down on my back to watch the sky. Dragonflies are dark circlings above my head, darting like visible fragments of the evening breeze. I turn on my belly and look at the ripples made by whirligig beetles and wonder what bodies of water there are near our new home in County Down. What ponds and lakes will I stare into there?

Sunday, 1 July

We have grasshoppers in the garden for the first time, springing up from the grass onto the arms of the swing chair, crackling in the heat. I watch one resting on the green metal and think about how amazing it is to have ears on your abdomen, tucked under wings – it's the tympanal membrane that vibrates in response to sound waves, enabling them to hear other grasshoppers singing. Each species sings in a different rhythm, so the female can mate with the correct species. I love how evolution finds these perfect systems and niches. I've never seen a grasshopper sit so still for this long. I give it my undivided attention. It starts stridulating, rubbing its hind legs against its wings, and the sound is loud so close up. I smile from ear to ear at this magic, and try to follow as it catapults into the air.

The grass in our garden is crunchy and straw-brown, the flowers blazing like a rainbow. The thought of leaving it all behind has been shadowing me for weeks. All morning I've been struggling to keep the anxiety beast at bay, and now I just can't stop the liquid panic rising. My heartbeats are manic. I can hardly breathe. This heat doesn't help. I reach for the sides of the seat and cling on, knuckles tightening. To stop it swinging, I plant my feet abruptly on the ground and feel a crunch on my sole. I know straight away it's the sound of a grasshopper dying. I am disgusted. I don't hear myself scream as a red mist falls, but can see Mum and Dad and Lorcan running outside towards me, almost in slow-motion, and I feel their arms around me, grappling, while the boom boom boom in my head says 'Whenever you try to do good things, bad stuff happens.' I have to fight back against the sweeping darkness. I know I must breathe. I know I can squeeze the nearest hand. I can sense sunlight but can't work out when I shut my eyes or how long they've been closed. The voices around me are meant to soothe, I know that. I know. But I'm submerged right now, completely under, and still babbling about digging up all the plants in the garden, 'I want to take them with us.' Someone answers, 'We'll try our best', which just isn't good enough. I open my eyes and feel drained and cold despite the heat of the day.

I stand and shuffle to the house where it suddenly hits

me, there and then: there's no school tomorrow. There's no school tomorrow or the next day, or the one after that, and now I can see all the evenings and days ahead stretch out without fear, without worry.

A wave hits me, and when I next exhale all the murkiness leaves and I can breathe again. I'm at home now. I feel giddy but can almost see the new feeling in the distance like the horizon, and when I think about the new home it is a much lighter thought because it also means there will be new places to explore, different landscapes, habitats, and all the new animals and plants in these places mean there will be no need to dig up the garden. What was I thinking?

I sit down on the step by the back door and notice the birdsong is less robust and gutsy. It lacks urgency. The work of spring and early summer is coming to an end. It happens, every year. I know this. The blackbird and all the other birds will sing again just as loudly next year. I've known it since I was a toddler, watching shadows from my parents' bed. The singing stops but always comes back. This realisation is close but still out of reach, too far away to be real. At least the swifts are still screaming, and they will be here for a good while yet. I breathe in the fragrances of dusk and notice flitting shapes moving in the new darkness – the bats are starting to come out to scoop up the midges. I close my eyes as a trickle of contentment passes through me. I'm pleased with myself for holding on today, for not letting the day end sour. I didn't let it swallow me completely. So here I am, enjoying the day into night, warm and still, bats replacing swifts in the cooling air.

there's loads of time ahead to rest and recover from school? (I'm so glad I live in Northern Ireland because we get July and August off school.) I want to go out adventuring and be busy, so I'm delighted with the consensus building. Then comes the inevitable debate about *where* to go, which is also when Dad appears in the kitchen to a storm of pleas and 'I wants' which ricochet off cereal packets and packing boxes, crashing back into the centre of the table.

Deep down, I don't really mind. I just feel unusually listless and want to get out, wherever we go. Apparently, Lorcan chose last time and it's my choice, so I suggest Big Dog. Even though I always choose it, we haven't seen any hen harriers there this season. I sit back and wait for the protest, and it comes like a hurricane from Lorcan, who wants to go wild swimming. And then Bláthnaid agrees, wild swimming is the thing to do, and I wait for the majority ruling to win. Strangely, though, it doesn't. Instead, Mum stands to grab a sheet of paper and starts writing a list of all the places and all the things everyone wants to do before we move house. 'Lorcan: wild swimming, Killykeeghan, kayaking, jetty jumping; Dara: Big Dog, hen harriers, Cuilcagh Mountain; Bláthnaid: pond dipping, Rossnowlagh Beach in Donegal, playing with friends at the park by the leisure centre.' Mum then assigns each of these activities a different day, so everybody feels like they've been heard and can say goodbye to their special places, and says that Dad and she can do the rest of the packing in the evenings, so there's time for full days. She makes it all sound so reasonable, and although there is a little more rambunctious conversation, we all agree that this is a great plan. Afterwards, things return to how we were: Lorcan is beating again, Bláthnaid's string is unravelling all over the table, Mum has completely missed *Woman's Hour* on the radio and breathes deeply as she walks off with Dad to get things ready for Big Dog.

We have a simple system to stop us arguing about music in the car: we all get a song each. And the cycle goes like this from youngest to oldest: Bláthnaid ('My Little Pony') to Lorcan (either Kygo or Motorhead) to me (punk), onto Mum (punk) and Dad (even more punk), which is great because, in a way, I get three choices!

Our journeys around Fermanagh usually take half an hour, which means two music cycles each – though Bláthnaid sometimes gets three, depending on traffic. Today is one of those days, so when 'My Little Pony' comes back on again Lorcan and I roll our eyes and try not to moan at the high-pitched rubbish about everyone being winners and other saccharine impossibilities. What a relief to arrive at Big Dog! We tumble out of the car with such enthusiasm that we offer to carry something to the lake.

It's around a fifteen-minute walk, through a Sitka plantation then into a piece of clear-felled forestry land with some newly planted trees. There are a few dead trees still, tall wooden spikes, which provide raptors with perching places. Although there are meadow pipits and stonechats flitting about, spiralling and clacking, I sometimes dislike it here. Maybe it's the barren feel of it all. If there weren't any hen harriers, I don't know if I would like it at all. In a few years, when the young saplings have grown into monotonous forestry once again, it won't be a suitable habitat for them anymore – hen harriers prefer willow and hazel copses. Still, it's gorgeous when you reach the top of the hill and the two shining lakes beckon in the distance – you just have to run to them, and we always do.

This time, about half-way down and mid-stretch, I stop because there are four alarming shapes at the edge of the lake – people! I know this sounds ludicrous, but here in Fermanagh we rarely see others, not in 'our' places – and having to deal with other people always makes me panic.

I calm myself and walk slowly towards the picnic bench. The rest of my family are still behind the hill so I sit behind a willow tree, hiding, catching my breath. I don't want to stare at the strangers; I distract myself by staring into the lake. Dragonflies whizz and skim the water's surface, their wings propelling them like bejewelled helicopters.

When the rest of my family arrive, Lorcan declares that he wants to go wild swimming here, right now, because he sees other people doing it. Mum and Dad discuss what to do. I notice the four shapes are now getting out, drying themselves and dressing, getting ready to leave. Maybe they feel the same way as me. Or maybe there's actually a relay of visitors like us, seeking isolation and a dose of wilderness, but we just usually miss each other and always assume we're all alone, all the time. We nod to greet the other family cordially, but as they disappear over the brow of the hill, wide grins pass between us all – we're by ourselves. Just how we like it.

The heat is like a furnace at the water's edge. While Dad heads back to the car to gather up towels and wetsuits (I've outgrown mine again, so I'll have to wear trunks), we make our way along the narrow grassy patch around the lake and lay out our things before dipping our feet in the cool water and eating our snacks. I lie back on the bank and look towards lines of Sitka spruce. It was two years ago when a pair of male hen harriers shot out of the trees like arrows, shouldering and sparkling against the purple heather. Glinting, rising, dancing, tumbling. I wonder if I'll ever see them here again. They've been absent most of the season and the place seems lifeless without them. I feel a dark, cold shadow creeping inside until a red damselfly distracts me.

Dad reappears with the swimming gear. I take my time, but Lorcan and Bláthnaid quickly change and rush in splashing. Perhaps they have freer souls. They're definitely more adventurous than me, reckless maybe. Or it could be my age: I'm more self-conscious now I'm older, more aware of myself. I still have vivid memories of being uninhibited like them, always talking, explaining, feeling intense, bubbling excitement. This early teenage phase in my life is quieter, more inward-looking, reticent, scarred by the hurt of others.

Watching Lorcan and Bláthnaid, I suddenly feel emboldened: I want to join in. I undress quickly and plunge into the depths. The cold hits me like an icy punch, I gasp, my skin tingles. I try and play with Bláthnaid and Lorcan but it just isn't working. I lie on my back instead and warm my front, squinting in the sunlight. I feel changed. I'm still changing. I lower my head further back into the water, turn around, take a deep breath and go under with my eyes wide open. I'm struck by the darkness, my chest contracts. The lake could be bottomless.

Doubt hounds me so much in life. If there's even the minutest chance that something might go wrong, that is still a number to me, a possibility. The desire to enjoy immersion comes with the fear of being underwater. Maybe others feel the same, I've just never asked them.

I come up for air, scramble to the edge, pull myself out. I lie back on the warm grass and feel the light and brightness all around me.

The horse flies (we call them 'cleggs') are out in force, the commandos of the fly kingdom. Silent strikers. They plague me, go for Mum and Dad too. It's a shame because they're such beautiful creatures. Beautiful but lethal. Eventually, we can't take it anymore. We decide to head to a local pub for dinner, so we can properly celebrate the first day of the summer holidays together.

There haven't been any hen harriers again today, but on our way back down the hill at least the ravens fly with us, and a grey wagtail too, bobbing among the rocks, almost invisible except for the flash of its lemon breast. I'm in good spirits, feeling easy. I skip, forgetting I'm a teenager. I run and laugh and shout, and we all run together and there it is, childhood, still hanging on.

Friday, 6 July

Walking is becoming my absolute favourite thing to do. I once loved to lie on the ground and wait for creatures to appear in front of me, but recently I've been far too brooding to sit still. I need to move.

Out on a stroll, our family are always a motley bunch. We can never control our excitement. We are gloriously uninhibited, and our progress is constantly interrupted by a leaf rustle, a flash of feather or a trundling dor beetle. It's wonderful to be together but I can't always phase out the chatter and flailing arms, the sound of running feet and shrieking laughter. The walks are lovely and maddening.

It's stop and start at Florencecourt this morning, as always. Lorcan and Bláthnaid are bounding but I'm finding it difficult to tune in. I slow down and lower my head, concentrate energy into looking. It always amazes me how Dad can talk, look and find all at once – I just can't do that. It's too much for me. I'd miss everything if I did. Lorcan falls behind with me, talking about his latest obsessions: video games (particularly the *Skyrim* soundtrack) and Soviet Communism. Today I welcome the talking and the distraction; it's a relief not to be so intent on looking and noticing. I still can't ignore the gleaming things when they appear, but to drift along like this also feels good.

Shaded by beech, birch and sycamore, we find it beautifully cool in the woods. The diffused light casts a glow all around us. Lorcan moves on and I settle into my stride, feeling the rhythm of my arms and legs start. I can sense the beginnings of a little musical, building with every step, until everything is part of the orchestra. The robins and blackbirds are strings. Great, coal and blue tits are wind, the corvids brass. The shrill cry of a buzzard is percussion. And they are all held together in time by the beat of my feet as I feel myself rising up and filling out and then. . . A shriek of discovery, not mine. It's Bláthnaid. I turn to see the widest smile on her face and a jay feather in her hand. Her whole being is shining. She's the queen of all feathery things and has been waiting for this moment for so long. She puts it in her hair and skips with elation. Mum takes some pictures: the girl and the jay feather in the late-afternoon light. We herd ourselves forwards again, carrying the warmth of the skies and Bláthnaid's find – when one of us finds something special, it replenishes us all. The same can be said about the way we share anguish, as we rediscover just a few moments later when a scream knifes the air. 'My feather!'

Bláthnaid's eyes are wide and filled with tears. It's gone. In her joy, the feather has fallen away.

We start to retrace our steps, dropping onto hands and knees occasionally to search the forest path. But the jewel is lost to us forever. I try to console Bláthnaid – the pain is real and all-consuming. She cries. She begins her meltdown – I know how it feels. I offer a piggyback home, scooping her up before she's had the chance to reply. The light bleeds from the sky as I sing her songs of nonsense. I feel her head on my shoulder, body relaxing. We keep going, keep pacing it out for ages judging by the twinge in my back, keep walking on until the need to skip takes over Bláthnaid again.

She slides off towards Mum, who puts an arm around her. 'You can have my jay feather,' Mum offers. 'The one from Scotland. We can write a story about what has happened with the photo I took.' Bláthnaid nods up to Mum and reaches for a hand.

Though we know the feather won't be rediscovered, we keep searching as we walk on the path, off the path, through the undergrowth, hoping for some other jewel to replace what we've all lost. And suddenly, there it is, a loud whirring crashing into the silence: a field cricket singing in the fading light. We all stop to listen. From a distance, we must look like a strange bunch, leaning towards a bramble bush. For us, though, the moment is holy. A tiny, solitary creature has the power to lift our spirits. A human catastrophe is transformed by a singing insect.

Saturday, 7 July

The bookshelves are empty. There are no photos or paintings on walls anymore. Our voices echo in the kitchen – there is emptiness everywhere, even at the height of the day's bustle.

My bedroom in the old garage is filled with packing boxes so we don't have to confront them in the house. It's not my room anymore. My posters and certificates have been taken off the walls, the periodic table is rolled up and my fossils, shells and skulls are all packed away with wings and feathers and sea glass. The space is there, but I'm gone from it. I don't want to be there anymore. And I'll have to get used to sharing a room with Lorcan now, because it will continue when we move to the new house.

I'm trying not to think about how hellish this will be. Sharing space. I will have to make allowances, we both will. We'll have to work out ways for us to compromise and have the peace and calm that we both need. It's not too bad at the moment, though. I enjoy the way the rooks gather on the roof above the room, their patter waking me up with a different dance each morning. There's also a robin that sings right outside the window – new palettes of sound.

We're gathered in the kitchen for breakfast, playing a constellation memory game, when Mum shouts 'Red squirrel!' Our chairs scrape the kitchen floor simultaneously as we push ourselves up and rush to the window. We don't see anything, save a lone blue tit at the bird feeder. Then an unfamiliar face breaks from the shadow of the trees, its small shape jerks forwards on the grass, stopping watching leaping, stopping watching leaping. There it is: a red squirrel. I stare in disbelief. To see it stray from the woods into this suburban place. I reach for my camera because no one will believe us. It's there in plain sight, bounding through our wildflower patch, scrambling up the trees and across the branches. An effortless acrobatic display, its russet body and exuberant balancing tail, swinging from tree to tree until it's out of sight. When everyone else is gone, I am still rooted to the floor.

Joy gives way to melancholy as I return to the echoes of the kitchen, the emptiness. In less than two weeks this will no longer be my home. New people will move in and they will not love it like we do. They just won't.

I go out and immediately feel how much cooler the air is this morning. I sit among the chattering fledglings and watch the hoverflies and bees feeding on catmint, ox-eye daisy, cow parsley. I breathe in all the memories and feel swollen with emotion. The greenfinches have just returned, alongside a charm of goldfinches. Flames of our mini forest, flames in our hearts. I feel an ache and lie down on the grass to watch the screeching swifts. My body sinks. I want nothing more than to sink underground.

Tuesday, 10 July

None of us can bear to be in the house anymore – its spaces are roaring out our intentions to belong elsewhere. It hurts, and it bangs against us as we move around. The urgency to visit all our special places becomes more intense. We have the list, we're working through it, we're running out of days. This morning we're off to Castle Caldwell Forest, carrying with us the bundle of memories from our different trips there, of wild swimming and caddis-fly larvae, chasing the call of a cuckoo, the newly emerged ringlet butterfly with its wings still coiled as it warms in the sun then flies away vibrating through the taller grasses.

It's turning into a blistering day, but we're cooled by a canopy of beech as we walk. Not native to Ireland, these Castle Caldwell beeches were introduced with other 'exotic' trees from the 1600s, during the Plantation of Ulster. Back then, County Fermanagh had many 'castles' strategically placed around Lough Erne – the Irish gentry were frightened of an invasion by the increasingly Puritan English Parliament and Scottish Covenanters, despite signs that everyone was starting to integrate. So these castles were in fact fortifications, and this one at Caldwell, built by Francis Blennerhassett of Norfolk, was originally known as Hassett's Fort. Although it was spared during the Irish Rebellion of 1641, while many other fortifications were burnt and their residents murdered, it has since decayed into the ruin we see here today.

Without going into history too much, let's just say that the events of the seventeenth century, between the native Irish and the new Scottish and English settlers, set off a chain reaction of ethnic violence that spilled out across the Irish Sea, sparking the English Civil War, prompting the execution of King Charles I, and leading to the rise of Oliver Cromwell – when Cromwell came to reconquer

Ireland in 1648, the Irish nobles were dispossessed and one-third of the population perished. The fault lines of this devastation are still felt near the surface of our shifting, uncertain world. We know all too well what little it takes to set us spiralling.

I try to imagine these ruins full of laughter and the sound of it being wiped away by war. Now, it has fallen into the hands of nature: resident cave spiders living in the cellar depths, roots growing and branches twisting and cradling birds' nests, red-squirrel dreys and bat roosts. I look up into the canopy, squinting, and back down to the spreading pools of light on the forest floor. From the castle walls comes the thrumming of bees, zipping out like an electrical charge as they busy themselves between stone crevices and the ivy flowers growing from the ruins.

We move on towards the wildflower meadow for our picnic. The meadowsweet here is glistening and abundant. There are cowslips and buttercups, too, glimpsed between the grasses like twinkling lights. I sit and inhale the honeyed scent. In Fermanagh, the boulder clay is so hard to drain and farm (thank goodness), which is why the meadowsweet thrives. In Dad's County Down – soon to be my County Down, too – the soil is so well drained that next summer we'll probably have to travel further to see meadowsweet. For now, it's right here in front of us, charging the heat of the air with sweetness.

When a ringlet butterfly lands on my shirt, I close my eyes to feel the flutter of its wings on my chest. My ears pick up more music: grasshopper stridulations, rook caws, invertebrate murmurings, the quaking grass and willowherb. A single song strikes out above the sound of everything else, three notes that rise gustily. I sit up and open my eyes, start scanning the trees with my binoculars. A lone chiffchaff is calling out at the top of a beech tree, its chest swelling and feathers rustling with the effort.

I look down at my top, but the ringlet isn't there anymore. It must have flown away with my sudden movement. I close my eyes and lie down again, wanting to feel the vibrations of all the creatures around me. I start to imagine myself covered with grasshoppers, butterflies, beetles, damselflies and hoverflies – they're resting everywhere on my arms, chest, face, hair, and because of their imaginary tickles on my skin, I laugh out loud until my eyes open and my body lurches upright, shaking itself suddenly and purposefully out of such a childish notion. And here it is again, my inner war rumbling on.

I'm still a kid really, but there's this piece of me that wants to be treated like an adult, and to behave like an adult. It's this 'maturing' self that for some reason starts worrying about what others think, and likes to pop bubbles and question the purity of these moments. But I'm in no mood for it today. Instead, I resist and lie back down to my reveries. No one can see, so who cares. No one can hear, no one can put me down or kick me in the face. I'm safe down here with the buttercups and meadowsweet.

I hear Mum call out. Over an hour has passed since I was last with them, apparently, so I make my way over to join the family. Passing the sentinel willowherb standing around the edges of the field, I stop to look for elephant hawkmoth caterpillars in the leaves, and watch the countless meadow brown butterflies feed, criss-crossing the luminous pink flowers. There's no tension in my body at all today. I am fluid and free. I hold my hand out and almost immediately a meadow brown lands. I am suspended in the moment, feeling hot sun on my back and the smell of meadowsweet filling my nostrils. I want this to be etched in me forever.

Friday, 13 July

Suburbia can feel claustrophobic. I don't know whether it is the place itself, the houses and roads and the people, or the views from the house. In Fermanagh, we're actually lucky because agriculture is still not as intense as it is in the east, but beyond the mown verges and roundabouts, the far away farmland we can see is all bright-green grass, square after square of it, with wire fences (where the hedgerows used to be), white fertiliser tanks, high-yield cattle, and some of it paid for by the state. All legal. All normal. Totally acceptable. The views are good, yet when you think about what's inside the view, all the wildlife it squeezes out, what we can see from the house begins to feel more grim and starts closing in. Which is why we seek out wilder places – places that aren't really wild, but feel like wilderness to us.

Today is cloudy but more refreshing, and we're leaving the farmland and monotonous greens behind by heading south-west from Sligo Road, up towards Marlbank Road where the limestone pavement rises from the grassland, with orchids and ox-eye daisy along the verges. As we near the entrance of Killykeeghan Nature Reserve, a ghostly shape glides past the car window and every McAnulty head turns to the left. There's a split-second silence before the joyous whoops of realisation that a male hen harrier just flew past, an unexpected messenger. I haven't seen one all summer yet there he goes, a talisman of delight, giver of silvery inner light.

The car fills with joy. We're all smiling right up to our eyebrows, and still glowing as we tumble out and run to follow its shape dip down into the willow trees. We stop and spontaneously hug each other – that's what we McAnultys do. We can't help it. We want to share our love and the joy we feel in a moment like this, share it with each other, with the place we're in. Mum squeezes us a little tighter and I almost feel like it will all burst out, all the grief I've been holding in, the darkness that keeps trying to pull me under. This is why Killykeeghan is also known to us as 'McAnulty chapel'. It's our place of peace and absolute joy.

Though we move off in different directions in search of more treasure, the skeins that invisibly bind us are spider-silk strong. Straight away, I see a dusty green-gold shape in the quaking grasses. Silently, stealthily, I move towards it and rest on a nearby stone. I watch as it opens and closes its veined wings, revealing ochre and night black. It's a dark green fritillary basking in the hazy sunshine. I watch it lift and glide effortlessly over the grasses. I touch the spot it has just left behind, to glean its warmth. I wonder if there are marsh fritillary butterflies here too, so I sit and wait for a little while, until the restlessness arrives again. I have to stand and keep walking.

In the brightening day I spy ragwort covered in cinnabar moth caterpillars – ragwort is a wildflower much reviled by farmers for its toxicity and danger to cattle and horses, but it's so beneficial for all pollinating insects. If you look closely at ragwort during the summer months you're sure to see the flowers vibrating with life, especially with the striped yellow-and-black caterpillars of the cinnabar moth, moving like slow-motion accordions up the stems.

Above me, a lone buzzard is keening and I turn to watch its wingspan fanning out and hovering over the field nearby. At my feet, the limestone pavement is cut with grooves and runnels, smoothed by water and time. In the gaps, alongside orchids and knapweed, devil's bit scabious is growing. The buzzard wheels over one of the luminous fields, that tedious green sea, searching, searching – and then suddenly drops, mantling its prey. That field just gave the buzzard food! I bow my head and smile: there is life in these fields, too. Nature is constantly surprising. Only by looking can we challenge our own prejudices, clearing them out and making way for possibilities. The sun breaks from the clouds and a single beam spotlights the buzzard with a halo. My skin flushes and tingles, I spontaneously jump in the air.

Wednesday, 25 July

We have moved. It's happened. We have moved county and house. I now live in Castlewellan, County Down, in a small, modern housing estate. We have native trees in the garden: rowan, ash, cherry and sycamore. Ivy covers the trunks and the forest park is right across the road. The last few days have been a whirlwind and now the darkness is so all-consuming, I haven't had the impetus to write. I hear the word 'depression' a lot but don't know if that's what I'm feeling right now, or if this is a normal reaction to the changes in my life. The effort of the everyday is like wading through treacle. Anxiety has been spiralling, and the energy spent on the battles is towering like the Mourne Mountains that now surround our home.

Last week, in between moving house, I did some filming with Chris Packham for his nationwide BioBlitz, which was assessing and recording the wildlife of fifty nature reserves across the UK. I filmed at Murlough Beach, just a ten-minute car drive from our new house, and I was so excited to explore a place that will become familiar. It was also the first time I'd worked on a group project. I usually work alone, but instead I was one of many young people presenting, which was easy really, because I was talking about what I love, what I feel passionately about. The problems started afterwards, when the comparisons and the comments started flooding in on social media. That's when my body started seething.

I really wasn't expecting such intense feelings of doubt. The words they used to congratulate or criticise me seemed to grow larger and larger on the screen, until it suddenly dawned on me that I sought attention and validation. This is something I hadn't experienced before. For so many years I've just done things that appeal to me, without thinking too much about them, and usually when I do them I'm on my own or with my family, isolated more or less from the eyes of the world – it isn't that I've been in my own bubble, just that not many people have looked in or cared about the things I'm up to. But with this, because I was with other young people, other activists and conservationists, I suddenly found myself obsessively comparing my words, my actions, even my face, with others. It greatly disturbed me.

If I'm doing this, then others must be, too. And where must all this comparing end? Surely the purpose gets lost along the way, as do we. The urgency of supporting collapsing ecosystems and protecting wildlife gets overtaken by human narcissism and insecurity. All this obsessing on Twitter last week has exacerbated my palpitations. There was nothing else to do except pull the plug, switch off. And I'm still switched off. But my enthusiasm and excitement are sullied. The words have hurt me, and I just want to hurt myself because of the shame and guilt and confusion of it.

I'm nobody: I've often heard other people say that to me, usually through clasped ears as I crouch on the ground. These words have been resonating for years, and for the first time I'm the one saying them. *You're a nobody*.

I do what I always do. What I must do. I went back to the dunes of Murlough Bay, with Mum, Lorcan and Bláthnaid, to be with the waves and the seals and the butterflies. I wandered along well-trodden paths listening to a linnet call, the skylark songs and the crying gulls. With every footstep, I tried to restore balance in my head, with my surroundings. This landscape – mountains, coast, sea, sand, forest – will shape the rest of my teenage years, and I must pay attention to it now, let my body be part of it.

Being autistic, I am a perfectionist and always looking for ways to prove that I'm actually an imposter, a failure. There are plenty of other people out there who fit the bill better than me, with large social-media followings, and who say the right things and look the right way, who stand up for wildlife and rage against climate change. I always believed I was standing up for something, and was beginning to feel that my voice was being heard. In my own way, I thought I was helping to fight for nature, by doing things locally, at my school, and contributing to science by recording data and taking part in protests. It isn't in my personality to go around regurgitating statistics about the horrors inflicted on the natural world, because they are outside of my experience. It fills me with despair and all I want to do is bury my head. Does that mean I'm weak? Am I too insipid? Does this mean I don't care? If it makes me switch off, why should others listen? I just don't fit in. I am not that sort of person, my body and mind just won't let me. I have to accept my limitations, or maybe my strengths. I'm hoping to help find solutions, but right now I feel like part of the problem.

I'm so glad we got out of the house, escaped the unopened boxes. I spotted a six-spot burnet moth on the devil's bit scabious, its red-and-black wings resting on purple, a clash of the Gothic with the royal. These tiny wild things lit up the overcast day at Murlough Bay, and as I lay on the sand, listening to the waves, I promised not to lose myself again. I must stop thinking about taking my own life. I don't want to imagine the world without me. I vow never to let it go this far, to talk to my family if I start holding on to the sadness.

The whole filming thing, the comparing, the validating, perhaps it's because there are deeper wounds. Maybe I use it all as an excuse. It's difficult to explain the weight the bullies have left over the years. I'm marked by them. I don't want to be. Without realising when it's happening, I am consumed, drawn under. It eats away my joy.

How can I overpower it?

How do I know they won't hurt me again?

To play my part in fighting for the natural world, I must start by smashing stereotypes. Every day I puff out an invisible black smoke, cooling and purifying my heart, trying profoundly hard to become me again. It will take time. I will have to be patient.

Wednesday, 1 August

I keep dreaming of Killykeeghan, the McAnulty 'chapel', back in County Fermanagh. My hands are touching limestone pavement, my feet pounding the earth. I wake up smelling the scent of Fermanagh but am not there.

I'm in my new room instead, and Lorcan is making music on his laptop, drowning the sound of traffic rushing in the distance. I fight back the tears that always arrive as I wake up these days. My brother notices that I'm awake and runs over. 'Pinch, punch, first day of the month,' he says. I growl back at him and hit out. He moves away muttering curses, confused by my reaction. I lie there, my insides popping, as a black mist rolls in.

A silky sound breezes through the window, and with every note it blows softly at the mist, puffing it away, until I can hear it clearly, almost hoarse, familiar yet strange. I rise and pull back the curtains to see a male blackbird hopping and pecking at the damp grass, extracting something delicious before bouncing over to the hedge. The young bird pops out from the undergrowth and the adult feeds it. I shift my body to watch more comfortably – the youngster follows its parent in a dance with three parts. Hop, peck the earth, feed, repeat. The young bird is incessant, vocalising its hunger with notes that already have a similar rhythm to the adult's.

We haven't put up our bird feeders yet – I remember the floods of tears as we took them down at our old house. The rain hadn't quite started yet, but I could smell it coming on the air. I was sat in the garden because the removal guys were in the house and I couldn't bear being near them. I tucked myself into the swing seat behind a low wall, where I pulled at the grass and let woodlice trickle up my hands. I watched a garden spider spin a thread of silk before scuttling behind a stone. I stood up slowly, looked at the back door. I saw the handle pushing down, Mum appeared. She came towards me and I can still feel her arms around me. I had been crying, and let a little more out with her. But I couldn't let it all out, there was just too much. I controlled it instead. Then it was time to leave, although the house wasn't empty. The removal company had brought the wrong van so there was still stuff everywhere. I could hear talk of what was going to happen next, but really it was just babble.

I can't really remember much else, and now we're here in County Down and Fermanagh feels so far away. I have to get on with my day, take each one as it comes. I'm so glad it's the school holidays. Imagine doing this and moving schools. Think of all those new people I'd have to process. Although, a strange thing has happened since we've arrived. There's a boy next door, a little younger than me, but he's interested in all things and likes to play board games. We've been sitting outside together, because it's been fine weather, and played cards and talked. I even showed him a colony of ants traversing the patio slabs, a marching line carrying crumbs and (amazingly) a small ground beetle. In that moment, I let my real self slip out. I was so excited that my mask fell. But he didn't laugh or look down at me in a weird way. Instead, he hunkered low and we shared the moment. The experience of looking with another person, somebody I don't know that well, was different. It was a little more shallow, if I'm honest. But the strangest thing was having company. This sort of encounter hardly ever happens to me. Afterwards, we carried on playing cards and chatting and I felt a spark of easiness that still feels good.

Later in the evening, we take Rosie into the Castlewellan Forest Park for a walk, which, amazingly, is fewer than three hundred steps away from our front door – even fewer if you hop over the back fence. Rosie is our constant companion on walks. A strong and silent guardian. She's pretty docile now, and obedient – a trait left over from her racing days, long forgotten except when there's a sudden noise like a gun crack or roaring car engine. We call her the 'autistic dog' because she always wants to walk the same route. If we're not all together, or if Mum isn't with us, Rosie stops suddenly, digs her heels in and absolutely refuses to walk. I remember once Dad phoning Mum while out on a lone walk, pleading for help because Rosie wouldn't budge. Mum had to go out with us in tow and physically move her. Since then, it's a standing joke that Mum is top dog. She-wolf.

The traffic is a little heavy on the road, but we escape it easily to enjoy the late-evening air. We couldn't do this at our old house, that's for sure. The busy road went on for miles before it reached Enniskillen and more busyness.

The walk inside Castlewellan is easy and I'm chatting with

Mum because I've promised myself, and her, that I won't hold things in to fester anymore. First I tell her how much I'm missing our Fermanagh places, and that everything here is so strange and different. 'It smells different,' I explain. 'Not in a bad way, it just does. It sounds different, too, in a good way. There are definitely more birds here, more insects.'

I then go on to tell her about Jude next door, my new friend. This makes her smile and the dimples in her cheeks become more pronounced – this happens when she's tired. There are also shadows under her eyes, and seeing them I want to find the beauty in everything and promise not to let the bullies weigh me down. I have so much love around me. I want to do it for her. I want to do it for myself. It's all around me, beauty, so why should it be hard?

Darkness comes in quick and it's time to head for home. We turn around at the lake and go back the way we came, the circular lake walk. Mum feels uneasy walking in the dark, in a new place, so we pick up pace and walk fast, breaking into a sprint across the road.

Nearing home, Mum grabs my arm and we stop in the falling darkness to watch shadows fly from one side of the road to the other. Bats. I see their flickers all around us more clearly than usual because the streetlight outside the back of our house isn't working. Mum and I laugh, and the excitement bubbles up. We rush back to the house: I find the bat detector and pummel through the kitchen and out the back door. In the garden more shapes mobilise from the trees – the bat detector is forgotten as I watch this origami take flight, just one shade brighter than the night, the bats' nimble wings making strange angles as they take to the air to feed. Each night we are blessed with these flying mammals, eating for us, too.

I stay out when Mum heads back inside the house,

watching the night sky. I notice a new feeling, a buzzing in the air, a pulsation that makes me look over to the buddleia growing in the garden. Something strange is stirring. It's whizzing with life and movement is palpitating in and around it. When the light goes on in the kitchen and I'm joined by everyone – Lorcan and Bláthnaid first, followed by Mum and Dad – I realise I must have shouted but don't remember doing it.

We watch in collective wonder as countless silver Y moths feast on the purple blooms. Some rest, drunk with nectar, before refilling, whirling and dancing in constant motion; even at rest their wings are quivering like leaves in a storm. The feather-like scales, brown flecked with silver, are shimmering with starry dust, protecting them from being eaten by our other nocturnal neighbours. I find it fascinating that silver Y fur can confuse the sonar readings of bats, and even when they are predated they can escape, leaving the bat with a mouthful of undressed scales. And here we all are, the McAnultys congregated in worship of these tiny migrants, perhaps the second generation. Soon they, too, will make the journey to their birthplace, silver stars crossing land and sea to North Africa.

The night crackles as the storm of flitting moves off, and even though the moths didn't make a sound, the night seems to have less noise without them. We jump up and down and hug each other, a stream of tension collectively leaking out and away and spreading. Let it go, let it all get caught in the night and taken far away. We chat and look at the sky, now empty of mammals, but still sparkling with Orion, Seven Sisters and The Plough. This is us, standing here. All the best part of us, and another moment etched in our memories, to be invited back and re-lived in conversations for years to come. Remember that night, when fluttering stars calmed a storm in all of us.

As I walk into the warmth of the house, for the first time I notice that there are no boxes. Everything has a place. The shelves are filled with books and paintings now cover the walls. This is our home, and like our house in Fermanagh, it will always be our home, even if we move on, because even if we flit again and again, this feeling will travel everywhere with us.

I give one of my little excited jumps and move my hands and fingers around with a happy yelp. Lorcan shouts out that I haven't done that in months.

'Are you happy again?' he asks.

'Yes!' I shout. I think I am. Am I?

Saturday, 4 August

No, as it turns out. A sense of being trapped returns as I wake up the next morning, and it stays there all day through playing cards with Jude and games of Gubs and Trivial Pursuit with the family, then on through eating pizza when even swallowing it down feels painful.

I tell Mum, as promised, about the stifling fabric on top of me. The invisible straightjacket. I just can't break free. Trains of thought are hurtling without meaning or direction. I'm stumbling from moment to moment, out of balance, out of sync, aimless and haphazard. Battling. Always battling.

Mum thinks that a new place to explore might bring in the cavalry to fight off these feelings. She also tells me that I need to hang on to grace and gratitude. 'Hold them close,' she says. 'And remember by writing down all the good things in life.' She's right of course, but it takes every single muscle to agree.

At last it's announced that we're all going out, despite

protestations from Bláthnaid who has quickly made a lot of new friends on our street and wants to play out. Extra guilt for me, because the point of getting out is to help Dara. Yet the frustration keeps building in the car. It's all so incredibly different from Fermanagh, and every car park in County Down is full. There are people everywhere. We drive from place to place without success and decide to go home again, but on the way back there's a space at Bloody Bridge – the name pays gruesome homage to the Protestants executed here during the 1641 rebellion, driven on to the rocks when an exchange with Catholic prisoners went horribly wrong.

Despite its grim history, or possibly because of it, the landscape here takes on a strange beauty. I can feel a breeze lift from the sea and cool the inshore heat, and with it the pummelling in my chest subsides as I listen for the rhythm of the waves crashing against the rocks.

We walk down over the steep stile and along the narrow path, rocks and sea on one side, dry heath on the other. We pause at a wider section and take in the view. Three men are fishing on the rocks – I can't help thinking what an idiotic thing to do, but perhaps it's just because more adrenaline is the last thing I need right now.

I sit down on the Silurian hornfels – the roughness of these rocks is softened by lichen and the thought that they're over 400 million years old, the result of colliding continents and marine life recovering from extinction. I stare at the granite veins, tracing them with my fingers. The coolness of the rock brings me comfort. Several wren chicks come skipping over the rocks, drowning each other out with attention-grabbing chirps. They pause, mouths open, their cries answered by diligent parents. I smile. I giggle. They come even closer, uninhibited by my hunched stillness, such a loud sound for a small bird. This is the music of our ancestors too, waves in one ear, wren siblings in the other. A two-track stereo. The sound of natural things that influence every other thing, whether we know it or not.

I go down towards the rockpools, where Bláthnaid and Lorcan have already taken their shoes off and are hopping like the wrens between the smooth Silurian crevices, stopping occasionally to crouch and look. I take my shoes off to join them, feeling the chill of granite. We stare into pools brimming with life. Hermit crabs scuttle between our submerged feet. I feel the tickle of a goby and blenny as beadlet anemones wave their antennae, scarlet with beads of blue around the inner edge. I touch one and a stickiness grips my skin – they do have sting cells, called nematocysts, but these can't penetrate human skin. The antennae retract, and so do I. But I'm firmly back inside the life I like to live, exploring, watching, learning. I start opening up too, tentacles of my chatter reaching out to Dad, sharing facts about the life we're witnessing. It just feels so good.

As the light fades and the day starts cooling, we put our shoes and socks back on. We turn back so that Bláthnaid still has time to play with her friends when we get home. Excited about seeing them, she runs out ahead but we all catch up when she stops and whoops. Her beady eyes have caught something again, this time the emerald sheen of a green tiger beetle. We pop it into a pot to watch it move for a little while, a glittering jewel, ferocious predator of ants and caterpillars. After ogling, we release it and watch as it javelins forwards, living up to its reputation of being one of the world's fastest insects. I bound up the steps, enjoying the weightlessness I feel. Will tomorrow be the same?

DARA McANULTY

Tuesday, 7 August

When we began, our feet trod lightly
Bare upon the earth, we were weightless
Travellers, allowing resurgence and
Regrowth, leaving enough.
Reverence.

Forging through millennia, we kept on
Adding endless weight, leadening
Heaviness, leaving deep and lasting
Indentations, sending shockwaves.
Eliminating.

Cruelty, cavernous greed, no impediment,
Hands and feet became Industrial.
Monsters, spewing toxicity, sickening,
Deafening, echoing arrows.
Piercing.

Now thundering, trampling boundlessly.
Decimating pathways once bountiful.
We watch helplessly, numb, aching,
Hollow, haunting cries in empty spaces.
Waiting.

Stop. I hear hope, purposely striding.
Footsteps pleading necessary action.
Great minds whirring, channelling change,
Demanding, respectfully our weight to
Lessen.

I want birdsong, abundant fluttering,
Humming, no more poison, destruction.
Growing for growth, it has to end.
Will my generation see the rightful
Rising?

Wednesday, 8 August

Every day we are exploring more of the forest park across the road from us, relishing it in small sections, getting to know it like a friend. We have found secret paths among the jays and rooks. We have climbed banks of leaf litter, strayed far off pathways. I can feel my energy returning, and my appetite too. I've not felt very hungry for days, but as the emptiness in my head is filled with fresh sights and sounds, the emptiness in my belly needs filling with food again.

The days are developing a pattern that we've probably all been craving. The topsy-turvy of moving and acclimatising is passing. We're settling into the house and into picnics in the forest. On one of the days, the day before yesterday, a hooded crow settled at my feet. It was a juvenile male, and I could hear the scratchiness of his movements as he hopped over my leg. It made me think of a line from *The Secret Garden*: 'Much more surprising things can happen to anyone, who, when a disagreeable or discouraged thought comes into his mind, just has the sense to remember in time and push it out by putting in an agreeable, determinedly courageous one. Two things cannot be in one place.'

Saturday, 11 August

We're driving to Dungonnell Reservoir, Glenariff, in the Glens of Antrim for an annual gathering of people from all over who care about hen harriers. It's a chance for us to express a shared outrage at the persecution of all birds of prey, and to share our own experiences of seeing hen harriers.

I haven't been around more than a small group of people for a while now, and I can feel huge knots tying up my insides. I start to fudge through it all with a fake smile, mumbling a few words here and there, until I find Dr Rooney (goshawk Eimear, as I like to call her). We talk about ospreys, red kites, drones, birds in general, and all of it flows so easily that it lifts my mood. Unfortunately, I can't talk with her all day and we're separated by others who want to say hello.

At all these gatherings, you see, well-meaning people tell me how inspirational I am. How my Tweets lift their day. How my blogs, campaigning, talks are 'just amazing' or 'fabulous', and some even say how I'm a 'fantastic role model to young people'. I hate it all. Honestly, I feel like an imposter. I don't deserve any praise. It makes me feel really uncomfortable because, well, why don't they just help their children, grandchildren, nieces or nephews to join in? To do the same. To take the spotlight off me.

I smile, shake hands. The usual.

I feel terrible not being appreciative of their compliments, and just have to walk away from everyone, down the grassy bank, towards the reservoir where the ground is scorched – wildflowers hang on by a raggedy thread. Dragonflies – hawkers – are hovering and darting over the boggy pools, snatching prey from the air. Peacock butterflies are in abundance – I count at least twelve that pepper the brown-green grass with fluttering colour and multiple eyes.

Back at the gathering, things are winding down. There aren't any talks this year, for which I am grateful – my normal enthusiasm for speaking to groups of people has completely vanished. Maybe it will return in time, maybe not. The rest of the afternoon passes with no big catastrophes, and we spot five buzzards on the way home. There were no hen harriers at the reservoir, though. Nor any on the way home, and I wonder if I'll ever see one again this year.

Arriving home to a welcome committee of Bláthnaid's friends is now a normality. I loiter outside for a bit, then all of a sudden feel the urge to go looking for some nature finds with the younger ones. There's a communal shrubbery across from our house, so I walk around it looking for a feather or some herb robert to show them. I don't want to bring out my own specimens, just in case they slip from smaller fingers or wander off. I notice a bloody feathered mass on the ground on the other side of the shrubbery. Perfect!

I run to get some gloves and take apart the prize: a goldfinch wing. I clean it off a little, quickly smooth down the feathers. I show the kids and they look at me and the wing with a mixture of repulsion and curiosity. I put it down for them to inspect, glorious and golden and black with silvery flecks of fluff. I tell them to stroke it, to feel how soft it is. They don't balk. Their eyes shine. I impart a few facts, and as some of the kids know Irish I tell them that goldfinches are called lasair choille, flame of the forest, and ask if they know that a gathering of goldfinches is described as a 'charm'. They ask more questions, I get my book and show them some pictures of garden birds. Who would've thought that looking under a shrubbery in a housing estate could have brought such a moment? I glow in the dusk. The streetlight is flickering, and a robin sings to it. I sit on the step, as the paths and streets are now empty. I wonder if there is a glow around me still, or if anyone can see it.

Monday, 13 August

The patio doors in the kitchen are wide open in the heat. I'm sat on the step playing card games with Jude as the sound of birdsong rises above the white noise of rushing traffic. We chat aimlessly about mythology and animals and – just stuff. I've never been good at conversation. It's an art I don't know the rules of. Either I just ramble on, spouting facts, not listening to the other person, or else I silently gawp, muddled by how to take part. It's how it has always been. With Jude, though, it feels easy. There's no third person, no chiding, no group, no bullies. I'm cautious, though. It's like I'm waiting for the contempt to slip out, even accidentally. It doesn't help that Mum is in the kitchen making plans to visit our new school for next week, which fills me with dread and anticipation all at once. The chance of a new start comes with the thought that I won't know anyone and haven't really met anyone outside of where our house is, except for Jude. I haven't really wanted to meet anyone else.

When he heads back to his house for lunch, a soft breeze picks up and blows eyed wings to my feet: a floundering peacock butterfly. I rush to get some sugar water, but it doesn't respond. I hold it up to the sky on my finger and it gives a little flutter. I pop it onto a buddleia flower, and it drinks a little. I wait and watch but it falls onto the ground. The end of a life.

I remember back to last August when Bláthnaid found a paper-thin, dusty peacock on the road, its wings still beating. She carried it home on her chest like a living brooch, and it stayed there all day as she spoke in gentle whispers and made offerings of food and water. When the end came, she placed it in her 'box of things', a memorial box of the once living. Although everything in the box is dead, they are alive in Bláthnaid's memory. She loves them all.

Sitting on the patio step, thinking about Bláthnaid's box, I feel a tear slither down my cheek. There's no living hierarchy in Bláthnaid's eyes, and therefore there is no living hierarchy in the world, not really. The smallest creatures carry the same importance and demand as much attention and awe as the ones that roam the savannah, fly through the skies or swing from trees. To Bláthnaid, to me, they are all equal.

Tuesday, 14 August

The shrieks of playing children swivel from house to house. Through the back window comes the intermittent rush of cars and lorries – it's not horrendous, though, because the trees in our garden shield us from the road. It's the first time I've lived in a house with mature native trees. The ivy-coated trunks vibrate with life.

Before breakfast, I usually leave the sound of Lorcan tinkling on his keyboard in our bedroom to see what's happening at the bottom of the garden, bursting with all sorts of wonderful things. And now that Dad has slung a hammock between the cherry and rowan, it's become part of my morning routine to swing on it before the rush-hour traffic gets too much. From it, I can watch a great tit feeding its young, flying off intermittently to forage for caterpillars and spiders. The young are fluffed up at the moment, and as dull as their exhausted parents. The feathers have a herringbone pattern to them, delicate and wispy, a barely-there green. Their calls (four shrill beeps) are answered quickly. It feels late for the baby season. It's typical for great tits to raise two broods, but having left our fledglings in Fermanagh some time ago, I'm not sure whether these County Down birds are on the first or second round of chicks. I need to tune in. It'll take time, but soon enough the seasons will tell me what I need to know. The turning of the year will reveal its secrets.

I close my eyes and listen closer to the four-beat food call until it's drowned out by a robin, cascading more elaborately in the humid air. A rustling of leaves alerts me to a young robin, so unlike its adult self – no red breast, a tweed body with ten shades of brown and a speckled crown. It hops to the right of me, in and out of the shrubs. On closer inspection, I can see it has actually lost the baby-white at the edge of its mouth and its feathers are sleeker, just starting to show hints of red. It hops with purpose and flies up onto our bird feeder. Our first visitor! We've had the feeders up for a week, but nothing until now. An adult swoops in with authority, the young robin scampers away to the cypress hedge and is gone. The adult puffs out its chest, poses and calls out beautifully, performing its act of defiance.

We all have a place in this world, our small corner. And we must notice it, tend to it with grace and compassion. Maybe this could be mine, this little corner of County Down, where I can think thoughts, watch birds, and swing gently on a hammock. But is this enough? Is noticing an act of resistance, a rebellion? I don't know but smile anyway because with each passing day I am feeling lighter.

Thursday, 16 August

Our garden is thrumming with birds today: coal, blue and great tits, blackbird, thrush, magpie, jackdaw, rook, all of them revelling on the grass, pecking at the feeders. I could happily watch them all day but rain is coming from the east, so we decide to go west to the coast at Murlough to stay in the sunshine. I usually despise being in the sun. I find its light too bright, its heat too hot, and it can make me feel as if there's nowhere to hide. But with this rain coming, I feel like being on the dunes at Murlough, pressed by warmth and sea breezes.

The ancient dune system here is six thousand years old, fragile and spectacular. The unusually high dunes were formed in the late thirteenth and fourteenth centuries by massive storms, and were used by people in the Middle Ages as rabbit warrens to provide them with meat and pelts. Grazing rabbits encouraged the grassy heathland here, but when myxomatosis first broke out in the 1950s, as it did in other parts of Ireland and Britain, the population was almost wiped out. With fewer rabbits, sea buckthorn and sycamore could grow, turning heath to scrubland. Now the National Trust have intervened at Murlough, managing the landscape so it reverts back to heath – and judging from the copious droppings, it looks like the rabbits are thriving again, too.

The day is sparkling, the wind is shaking and shaping the clouds. Lorcan and Bláthnaid want to swim, so I stroll along the beach with my binoculars. Shapes out at sea stop me: a trinity of torpedoing gannets. They swoop, wheel and suddenly drop, spiralling until the last second when they transform into arrows before hitting the water. Swallows are overhead – I can see their small bodies so clearly, weightless and constantly moving. I feel myself rising with them.

My dark, knotted thoughts seem to be staying away at the moment. I feel as free as the gannets and swallows. If they can live their lives, shouldn't I do the same? Can I breathe and live and also fight? The natural world – which includes us – is facing such enormous challenges that it's easy to become overwhelmed and depressed. But we must fix them, and if I'm no longer here, alive, I can't be part of the solution. What is it that's holding me back? Anxiety? Depression? Autism? These are the shackles. Surely, I can break free. Or at least I can accept these things as part of me. I have no answers, but the lightness of these thoughts and these days weave my body and mind with everything around me. The only thing that I am really bound to is nature – as we all are.

Lorcan and Bláthnaid are running towards me, and I run to them and then we run together, exultant. We slow down in unison, all feeling the same pull of the large peculiar shells dotted around the beach. We each pick one up and hold them outstretched, delicate porcelain in our hands. They look like pale planets, pockmarked with symmetrical lines. I rattle mine, listening to the whisper of sand and the past. These are sea potatoes – a type of sand-burrowing sea urchin – with pocks that once held spines and a bleached calcium carbonate shell that could so easily shatter on land or at sea. Each one is a miracle. So many miracles washed up at once.

We start to collect them and Lorcan decides to name three of the best specimens: 'Sandy, Sam and Sandra'. He holds a conversation with them, three sea potatoes, which makes us laugh so hard that tears well up and almost flow, and we're still laughing as a warm rain starts to fall on us. Under dark skies, I feel completely unburdened of any doubts in my own abilities to help our planet. Instead, I feel energised and ready. Sopping wet and cold and with chattering teeth, still giggling madly, I feel hope pouring in the rain. Being myself is enough.

Sunday, 19 August

The air today is sweet on the tongue. For days, I have been seeing everything like Dorothy in Oz. I'm not quite sure what has happened. Maybe the serotonin levels in my brain have miraculously reached a point of equilibrium. Maybe talking to Mum and writing everything down have helped. I just don't know. The mist has gone and I can see all the fine details.

This morning Dad is driving us all to Tollymore Forest, one of the first public parks in Northern Ireland, opened in 1955. The rain has stopped, the intense heat of previous weeks has gone. Before I get into the car, a funny sensation prickles over me: there's a small creature on my shoulder.

It takes me a few seconds to realise it's a water boatman, unrecognisable and naked out of water. I ask Dad for a positive identification and we all marvel at this sumptuous creature. The oar-shaped hind legs are still outstretched, resting on my bright-blue fleece as if it were the surface of a pond. If I hadn't felt it, we might have been deprived of a magical moment – and it is these, tiniest of noticings, that bind all of us together. Nature's miracles. The water boatman starts warming its wings and flies off, disappears from sight, but leaves us all with the gift of a conversation that lasts until we reach Tollymore Forest.

There's an overwhelming amount of people and noise in the car park when we arrive, which reminds us why we've not visited here yet. This onslaught on my senses fills me with dread. I try to push the thoughts aside, distracting myself with the large route maps. We decide to walk the second-longest 'red' trail, not too strenuous but hopefully less busy. As we enter the forest, the crowds do begin to thin and birdsong overtakes the human chatter.

Usually, our family walks are very slow but today we yomp purposefully along the Shimna River, crossing Parnell Bridge, to leave the throng behind us. A patch of gold catches my eye: a stagshorn fungus, tendrils of basidiocarps snaking from the ground. It's spongy to touch, slightly squidgy. Beautiful in its radiance, a sun lamp on the forest floor. I rummage around and find the piece of wood it's growing from, covered by leaf litter and surrounded by luminous moss. Its Latin genus, *Calocera viscosa*, means 'beautiful and waxy' (*Calocera*) and 'viscous/sticky' (*viscosa*), though it doesn't feel that sticky right now because the rain the other day was short-lived and it's been dry ever since.

Tollymore began to be planted as an arboretum in 1752, with a mixture of native trees and exotics like eucalyptus and monkey puzzle. The oak from Tollymore was used to fit the interiors of White Star Liners, including *Titanic.* We speed through it all, trekking up onto higher ground where I stop to listen to a buzzard and catch a glimpse as it dips behind the trees. Further on, I bend down to tie my bootlace and see in front of me something discarded but utterly beautiful: a nest. I gently pick it up and turn it around in my hands, enjoying how intricately woven it is, with twigs, roots and moss, while the inside is still layered with hair and feathers. My mind wanders through the possibilities of why the nest was on the ground: had it been raided? Did the wind blow it down? Was it tossed from the tree after the broods had fledged?

I carry the nest with me as we walk, in awe of the complexity, the craft. A scuttling shape emerges: a spider, a garden cross, with the cross shape and spots of white on its abdomen. I love spiders, especially the garden cross and orb-weaver. They're such a beguiling sight – it hurts me to think how people so carelessly kill them or just remark on how disgusting they are. When the garden cross scampers back into hiding, I put the nest back onto the forest floor, even though I really want to keep it. It may no longer be used by birds, but it has become a shelter for the spider, and possibly a source of food. A pocket-sized habitat that I don't want to disturb.

I'm quite a long way behind the others, so I rush on to catch up, and even skip a little because I feel so lucky to be a part of my family. By the time we reach Hoare's Bridge and the trickling of the Spinkwee River, we've walked quite high and can overlook the jackdaws and rooks congregating in the trees below, a parliamentary meeting that probably has more interesting ideas than our human governments seem to have.

The more I read or hear about politics, the stronger my reaction is to focus on nature and wildlife. Just thinking about our situation here in Northern Ireland gives way to intense anger and frustration – the two main parties sticking either side of an old divide. Do I have to be inside Stormont to make a difference? Is it all about Westminster or the United Nations? Can I fight for change from the outside?

I listen to the crows again, and allow their sounds to go deep, to that place where memories are stored. I can hear the buzzard mewing too, but can't see it. I close my eyes instead, to rest a little, and so I can listen to the gentle rush of the river. A blackbird sings – maybe its last song of summer.

I carry on, running downhill until Altavaddy Bridge, where the Spinkwee converges with the Shimna. The water gushes over rocks. The banks are sprouting with damp tree roots that almost touch the river. Lorcan and Bláthnaid have taken their shoes and socks off already and wade through the water. I sit on the edge where a dor beetle ambles onto my trouser leg – I notice the bluish sheen of its legs and the gleam of its coal-black elytra. I lift it up and turn it over with my thumb on the palm of my hand. These luminous and exquisite creatures are the true cleaners of our countryside, consuming their own weight in dung every day. They also have amazing mating habits: after sunset, the pair find a suitable cowpat where the female burrows to create chambers; the male works behind her, clearing and depositing another parcel of dung in every chamber before the female lays an egg in each – when the egg hatches, there's a ready meal waiting for them. Life cycles like this make me so happy! The beauty and logic of it all.

I'm still lost in dor-beetle mating rituals when I hear shouts from Bláthnaid, who has managed to slip on a rock and get herself soaking wet. Lorcan is sopping too – he obviously favoured a dip with his clothes on. Unprepared, Mum gives them her jumper and the two of them squelch together back to the car park.

Wednesday, 29 August

You know it's blackberry time when Seamus Heaney's words start to echo around our house:

> Like thickened wine: summer's blood was in it
> Leaving stains upon the tongue and lust for
> Picking

We've spent the morning gathering from the roadsides and in the forest. Tasting my first blackberry always sends a little spark deep inside. A glowing sweet fire. As the juice dribbles down my chin, I feel that freedom again and ride the electrically charged thought that all things, good or bad, have an ending. And by the time I've had a handful, I even feel a bit better about the way I completely blanked the principal when we visited my new school yesterday.

He had started talking about being a punk in the 1970s because he saw I was wearing an Undertones T-shirt. I should've been overjoyed to have something in common with him. Instead, my brain didn't play along. My head throbbed. My eyes and ears couldn't process anything. My stomach was churning and there was a foul taste in my mouth. But thankfully it started to fade as we walked around the school. I probably relaxed because the vice principal seemed to have a sixth sense – she gave Lorcan and I plenty of time to take in our surroundings. But the stress of starting again was visceral. Maybe it was made stranger because the school is a mirror image of my old school: they were built in the 1990s with what must have been identical architectural plans. There's no logic I can apply to overcome these waves of feeling. My senses, my body, my system, just won't let me.

Returning home from the forest, I take to my favourite place: the hammock. The air is cooler now, the garden quieter (well, besides the traffic). The shadows are lengthening over the mountains where we've seen red kites wheeling high on thermals. The wings of our bird neighbours are still beating, and the swallows are still here, growing in number by day, feeding together, bubbling with the imminence of longhaul flight. Some swallow pairs may have had a third brood late in the summer, and even these fledglings are ready to join the adults in making the treacherous journey to southern Africa, via France, eastern Spain, Morocco, and either over the Sahara, around the west coast of Africa, or east down the Nile valley. This incredible migration will never cease to amaze and inspire me – that these small powerhouses can fly 200 miles every day for six weeks, in a race against starvation and exhaustion. When I start to worry about school and all of the newness – of people, of classrooms – I think about the resilience and determination of swallows.

AUTUMN

our interconnectedness, our interdependence.

But is writing enough? It doesn't feel like it's enough. Not nearly enough. I need to think of other ways of rising.

The leaves of the birch tree diffuse the light. The fly agarics are red jewels spotted with white flakes, flaming bright and triggering in me flashes, evocations: I'm four years old again, hunkered down, facing a man with long white hair and glasses – I have just started wearing glasses, too. He has a wooden box filled with fruits of the forest; each different fungus is fascinating, alluring, utterly mesmerising. I feel it, you see, the connection. Even back then, so I listened intently and relentlessly asked the man questions. His kindness isn't all remembered, but an impression of kindness was imprinted in me, and the spark was lit. A need for learning was burning.

I don't remember more of that day, if only I did. What did it smell like, what sounds did I hear? What did I really say? Mum and Dad have photos of me: tiny and serious and spectacled, and obviously interested enough to shake out my money box and go accompanied to Waterstones, where I laid out the coins on the counter for my first field guide, Roger Phillips' Mushrooms and Other Fungi of Britain and Europe. *Mum bought me picture books, too – my favourite was* The Mushroom Hunt *by Simon Frazer, with its glorious illustrations and wise words. Both guide and picture book are now worn, dog-eared, still loved.*

My body feels light as I roll over onto my stomach and stare at the agarics until they blur. Regarded by shamans as holy mushrooms, they were given as gifts on the winter solstice, perhaps because they are hallucinogenic (although deaths from this particular Amanita are rare, I wouldn't chance it). They are so beautiful, though. The quintessential fairytale toadstool. Some are rounded and small, just starting to blush. Others are more like gnome platters, bright and peeling. I touch the spongy surface, slightly damp and sticky. I smell it: only a faint odour of sweetness. I swivel around again and think of the season ahead. New beginnings in school – a school protected by mountains to the back and side, facing the sea on the other side. A new horizon. I vow to stand proud. I have a mission. I have a journey to make, doubtless with obstacles, but they won't stop me, just as you can't stop fruits bursting from tree and soil. I can battle quietly or loudly, with humility. I can be rooted, to ideas, plans, hope. I can grow. Sapling stage is ending, it is time to grow thicker branches and mature.

Sunday, 2 September

I've taken to walking to the forest park most mornings. There's a spot I've found just behind the peace maze, on a patch of grass off the main path, where I can perch unseen facing the frothing willowherb, its feathered seeds blowing away in the breeze. From here I gaze at the mountains on the horizon, or watch the rabbits run in and out of the warrens – sometimes they come right up to my stillness, about twenty of them, all twitching noses and scampering shapes.

The horizon at my old house at County Fermanagh was Cuilcagh Mountain, as flat and inviting as an outstretched palm. A plateau of protection. Now we share space with the Mourne Mountains, rearing up with textured undulations, valleys and peaks, our Narnia. I long to run into the crevices and along the jagged edges. As time goes on, the Mournes and I will inhabit each other.

Here, the traffic doesn't purr as loudly as it does in our garden. I lie back and watch the jackdaws as the heat of the day starts to rise. The birds cavort playfully, their space-invader sound effects reverberating against the forest floor. I feel the ground move, as I often do. I feel all the movement down below, all that life. And it's in me, too. As I head back to the house, pausing by a huge patch of willowherb, I can hear the grasshoppers still singing.

The house is busy. School starts next Tuesday and uniforms for Lorcan and me hang on the kitchen door, like a taunt. The emptiness of the garments, hanging loosely, waiting for me to fill them; I wonder if they will get torn by others. I move stiffly towards the throng in the kitchen, making sure not to brush any nearer before I need to. The map is on the table, and there is a strong smell of coffee. Bláthnaid is pressing dandelions into a notebook, Lorcan is reading an Usborne history encyclopaedia. He's still fascinated with communism and the Cold War – hammer and sickles drawn on paper surround him.

When we (by 'we' I mean autistics) get interested in something, most people would call it an 'obsession'. It really is not an obsession, though. It's not dangerous, quite the opposite. It's liberating and essential to the workings of my brain. It calms and soothes: gathering information, finding patterns, sequencing and sorting out is a muscle I must flex. I prefer the word passion. Yes! And it's absolutely essential that we get to follow our passions.

Our feet are itchy, the desire to head out into the day is constant. The warmth beckons, so we head to Crocknafeola wood for a good walk – not a mountain walk because we've things to do later, and mountains are for the stopping of time; you need hours. The peaked guardians are everywhere, though, and I can feel Slieve Muck on my back as I look towards the path ahead, like a lonely giant that somehow seems apart from the other peaks. We pause in the car park to hoard blackberries, stuffing ourselves among the coconut-scented gorse. We follow a dirt path and the call of a stonechat uphill, keeping to the edge of the forest. It's mostly plantation here, but there are patches of willow and hazel breaking through.

As we move away from the gaze of Slieve Muck, I can feel my feet treading more lightly and my heart rate start to slow – my anxiety about school flowing into the earth. Then I feel the sizzling anticipation that something is waiting for me, and as I glance down there's an orange fluttering, a gauzy light sprinkled over amber wings: small copper butterflies, about ten of them, communing. Some are ragged, others pristine. They flit and rest on each other, those with worn wings and those with still velvety and bright wings, journeys beginning and ending, all as one.

I reluctantly leave the glow of the butterflies behind as we hike up and onwards around the forest, where clouds of midges compete for sunlight in the cool breath of the trees. The walk becomes monotonous but the light remains extraordinary, turning the pathways golden either side of plantation darkness. Dragonflies zoom overhead. The rattle of jays casts a spell. I continue to feel light-footed until we reach a section of pathway which is flooded – either we'll have to make our way round it by walking up a bank tangled in brambles and gorse, or go straight through, wading. Lorcan and Bláthnaid are already taking off their socks and shoes, laughing, almost hysterically excited. Dad realises he'll have to do the same. Rosie can't manage it – greyhounds, especially if they've had a tough beginning in the racing trade, can age very quickly. She daintily and distastefully shakes her paws if they get wet – maybe we've pampered her for the last five years, but she's not an adventurous dog. Mum asks if we need help and receives shoes and socks gracefully then turns down the opportunity to wade herself, and sets off through the dense growth of the verge. The sensory feeling of mud should be wonderful for a young naturalist, but it's something I'm still learning to enjoy – I don't know why the squelching is so excruciating. I choose the hard-earth option, even though scratches and cuts are inevitable.

I'm distracted easily and find a bilberry bush with five seven-spot ladybirds basking in the sun: one opens its cloak and, with humming wings, flies the short distance to my outstretched finger. It rests there for some seconds before ambling down onto my wrist, where I feel the tickle of its legs. It flies away when a beam of sunlight lands on its body. I stay still, watching the remaining beetles, the way the shadows rise and fall, the brightness of the red changing with the clouds.

The path is flooded for some distance on and off – there's been a lot of tree felling recently, which leaves me wondering if that's the reason. The pools are peaty, some have rainbow slicks. Bláthnaid is leading the way with water up to her calves, clutching her puffin teddy from Rathlin Island to her chest. Dad and Lorcan follow with the ever-reluctant Rosie. Lorcan is the official Rosie whisperer. Their bond is the strongest. You can see it in the way he usually coaxes her through the puddles. But today, at the end of the pathway, although the humans are in good spirits, poor Rosie is shaking out her paws in disgust, looking very put out by it all. Dad's knees are wet too, because he decided to lean right down to look at a whizzing great diving beetle. I remember the time one flew into our bird bath in Fermanagh, and how we marvelled at this opportunist with an air bubble on its back, carrying its own oxygen supply wherever it went, a predator on the prowl, eating everything and anything. We turn the corner and emerge out of the forest, facing Slieve Muck, back where we started. We wipe our feet as much as possible before making the short descent back down to the car park. It's so hot, now the sun is at its highest, and there's still so much of the day left.

On the way home from Crocknafeola we stop on the bridge over Whitewater, just outside Kilkeel, which is a particularly lovely stretch, fast-flowing with a rocky weir and the highest rocks covered in moss and dripping with river spray. The salmon here wriggle and jump over the weir before resting in the pools above and travelling on and up. Hawthorn and alder overhang the river, lime-green leaves almost touching the water.

Our family is well used to finding itself at destinations chosen for their proximity to freshwater. Since moving to County Down, Dad has been photographing all the rivers of the Mourne area in the south – he did this in Fermanagh too, finding the sources, their stories, and how they overlap with the language and culture of the places they run through. I'm staring into the weir when a bobbing movement shows itself: a white-throated dipper – current-swimmer, rock-clinger, water ouzel and the holy grail of stream-watching. It hops from rock to rock before disappearing, all in the space of Dad's camera click.

Saturday, 15 September

The first fallen leaves are pirouetting at my feet, rising, tumbling, skittering, falling again. The breeze feels colder with the emerging autumn. I'm standing on the foothills of Slieve Donard, mother of the Mournes, towering above the other peaks that clamour around like children at her heels, desperate to learn how she grew so tall. The Glen River is gushing and drowns out the sound of Lorcan and Bláthnaid climbing trees. I sit and watch the white water swells, tinged with earthy brown. I feel like a speck of dust among this expanse of forest shadowed by mountain. Deep clumps of knotted oak roots spread out, a staircase up and up to an unseen pinnacle chafed by walkers, so many walkers. I nestle out of sight under an alder tree, seeking out a pinch of twilight against the brightness and noise of the day. The sound of this gushing mountain river is not dissimilar to the clamour that has been in my brain all week.

What a week it's been. I woke up in a pool of sweat in the early hours of Monday morning. My heart was racing and my chest was so tight I thought I'd choke. When it was time to leave the house, taking those first steps away from the front door, my whole body was rigid. Then in the car park, after we had said our goodbyes, I issued computed responses to Dad so he wouldn't worry. 'Thank you. Yes. It'll be great. I'm okay.'

Fear constricted all and everything as I walked from the car with Lorcan to the school gates, with the teenage shrieks and chatter taking on an arena-sized volume in my head. I stopped as we walked, paused to look across the football field at the twenty or so oystercatchers waddling in peace, probing the earth for worms. I felt old wounds opening. Lorcan tugged at my blazer for me to move on but I shrugged him off for just a moment more. I needed to take in the black-and-white plumage, the beaks like orange spears stabbing the ground. The oystercatchers started piping, whistling and trilling – nobody took any notice, which also meant no stones were thrown and the birds were left alone. The noise grew and grew, and without encouragement or disturbance they lifted and flew up and over the trees and houses towards Newcastle beach. I looked up after them, to the blue sky, twisting my head to see the high peak of Slieve Donard, and found it miraculous to think that my school would be at the foot of the tallest mountain in Northern Ireland, one of the twelve chieftain mountains on the island of Ireland. I felt embraced by it. Slieve Donard would be with me everyday. A warmth flooded my body, untying lots of little knots.

Lorcan and I were met by Karen, the vice principal, who guided us towards the sports hall. Some of the students wore hoodies and sports tops with the school crest on it, others were in blazers like us. There was a bubble of anticipation around everything but neither of us knew what to expect, what rules and regulations to follow.

Lorcan was met by his 'buddy', and I was introduced to mine: Felix. Although it was weird and uncomfortable to start with, when we started talking there was lots we had in common, like our love of science and maths. I could already feel the tiniest spark of friendliness as the day sped past with new faces flashing by. There were students from Canada and the Isle of Man, which meant Lorcan and I weren't the only new kids. But there's nothing like finding somebody similar to yourself, who enjoys the challenge of having their intellect stretched.

In my last school we had a board- and card-game club and although we didn't talk much (we were all on the autistic spectrum), there was a camaraderie that was a lifeline in such a challenging environment. Many of us wouldn't dare go outside. When we did, we were immediately the bullseye on the target. It was as if we wore bright-neon beacons that basically said yeah, come and beat me up for being different.

Would it be the same here?

Having a sounding board (and a human map) in Felix helped me navigate the rest of the week, as I flowed easily from class to class, enjoying the lessons and spending my breaks and lunch walking around the school grounds debating with him. I had never talked so much at school, ever. I must've spoken many thousands of words more in this single week than I've uttered in my entire school life so far. Discussing science, *Star Wars*, nature, maths, philosophy, history. Everything. I even started to wonder if this was what being normal actually felt like, but had to stop myself because normal is definitely not something I want to be. It felt strange, alien. But such a relief.

Above the sound of the rushing River Glen, I hear my

name being called. I've been hidden below this alder, out of sight, and the torrent of worried voices from Mum and Lorcan tells me I've been there a while. I get up to join them, stepping into a pool of sunlight where some serious tree climbing is going on. I stare back to the river as a grey wagtail bows to the rocks, disappearing into the undergrowth like a river nymph.

Wednesday, 19 September

This morning, as Lorcan and I walked to the bus stop, we saw the damage of last night's raging winds: trees toppled over, branches brutally snapped. Some had escaped their prison of decorative concrete or clay pots. One tree, an oak, growing below the pavement had fallen to expose its root ball, so tight and tangled that there couldn't possibly have been any more space for life. It wasn't the wind that toppled the oak, not really. Being confined in asphalt and under slabs, that's what did it. When we strolled past on the way to school there were traffic cones all around it, but I stepped inside the space anyway and wondered if anyone saw me touch the bark. 'Sorry,' I said.

The ripped-up human surfaces, all broken and jagged, spoke of people first, nature last. I knelt beside the trunk and stroked the bark, no longer caring if all the people passing were watching or not. I pulled some still-green leaves from its branches: they were still perfect. I collected a handful of acorns from the branches and put each one into my pocket, like small pieces of hope. I walked on with heaviness, but knew my blazer carried something good.

In the afternoon, when we got home from school, we planted out each one. They may or may not make it, but fifty-fifty is enough and we should always take the chance.

As I write at the end of the day, I press the oakleaves into my diary in good company with feathers and celandine, gentian and speedwell.

An unfamiliar rhythm is beating, gentle yet raging. I have gone two weeks without being bullied. Two weeks. This is the longest period I've experienced without taunts and jibes or fists landing. It feels strange, almost eerie. I had prepared myself for the worst, because that's what I've come to expect. I had my list of affirmations and vault of memories from Rathlin Island or my garden back in Fermanagh. I had strategies worked out in my head, of what to do when things got bad. I had even written conversation openers to hand to my mum, if things got cloudy. Instead, every morning I walk and sit with rabbits and rooks, go to school, work hard, talk animatedly with my friend Felix as we watch the gulls and oystercatchers quarrelling, rising and resting. Then I come home, with energy to spare because I've not used it all up battling anxiety. I do my homework. Write my diary more and more. Watch birds. Play my computer games. It's weird, feeling ordinary. Usually, every breath of wind is a storm. At the moment, the wind is gentle and I find myself laughing when it swirls around me. I am happy, yes, but with it I also feel more cynical and hardened.

Over the years, a wall of stone and beautiful ivy has grown up around me, and only family and wildlife are allowed in. Although shafts of light are starting to get through all this, I am still wary and catch myself wondering how long it will last. This doubt creeps when the wall and the ivy are in shadow. But I'm starting to realise that I probably need both the light and the shadow. They are part of me, and I can't change that.

Friday, 21 September

My social media has been a hive of activity these last few weeks: the naturalist and TV presenter Chris Packham is organising a People's Walk For Wildlife in London, and has asked me to recite 'Anthropocene'. I call it a poem but I'm not sure it is. I feel it would be good to say aloud, to a crowd. I've only written a few 'poems' in the past, none of which were memorable, but with this one the words spilled out and I kind of 'performed' them, recorded and shared them on Twitter. *Bare upon the earth, we were weightless... Will my generation see the rightful, rising?* Lots of people liked it, including Chris. It's always a surprise to me, that people appreciate what I say and how I share it.

These past weeks I've been helping raise awareness for the walk in London by doing videos and Tweeting lots. It's an exciting prospect: hundreds, if not thousands of people marching on behalf of wildlife. I'm not worried about speaking. I actually find it easier if there are lots of people, because I don't have to make eye contact and it's much easier to blur them into a mass. Speaking to smaller groups, that's a killer: you feel the heat of their gaze, every twitch, each sigh. No, talking to lots of people is not something to be afraid of: all that space swallows me up.

So I have an early flight to London with Mum in the morning. I feel bad for flying, we both do, knowing the damage emissions do to our world. But we're not guzzlers. We're not jet-setters and never have been. We've only had one European holiday, to Italy, which was six years ago now – I still remember searching the shrubs outside our caravan and kneeling down on the dusty earth with the lizards, the heat unlike anything I'd ever experienced as I put a stick on the path and watched ants crawl up one by one. Though I don't mind going to other places, I prefer familiar things really. My parents have never flown much either, so I suppose we don't have lots of carbon in our past. Ideally, we should be taking a boat and driving to London, or taking the train, but it's beyond us financially right now, and I can't get more time off school so soon without getting into trouble. The walk feels like important work, something we should do.

I've already got the poem locked in my head now, *When we began, our feet trod lightly*. I know it off by heart. *We want birdsong, abundant fluttering, humming, no more poison, destruction.* I feel excited. Perhaps it is the right time for me. Tomorrow will be epic.

Saturday, 22 September

I'm sitting with Mum in our London hotel room, drying out our clothes and the contents of our rucksacks. My bones are getting cold now that the adrenaline and charge of the day are fading. And what a day. It's going to take a while to process it all. My body and brain are exhausted.

We arrived very early this morning at Hyde Park, straight from the airport. Thousands of people had already turned up, and below darkening clouds thousands more arrived on a day that blazed with human empathy and camaraderie. I saw the young campaigners and many others I've only 'met' on Twitter, and all the meetings and shaking of hands is still blurring my brain. Circuit boards are snapping.

I was soaked to the skin, hair dripping wet, like everybody else standing in the pouring rain. Anxiety started swirling. A silence seemed to descend on the crowd when I stood in front of them, an expectancy. But I felt strong on the stage as I spoke. My words were purposeful, filled with fire – I hope I've managed to ignite others.

I ad-libbed a lot at the end, and can't remember exactly

what I said. All those frustrations that so often leave me feeling helpless started pouring out. All the times I've spoken to people who didn't listen or didn't care. For all those brick walls and slammed doors. I poured out my feelings, passed them on. Who knows if my words will help.

The speeches after were all magnificent. Intergenerational. Important and inspiring. Afterwards, as we walked from Hyde Park towards Whitehall, we played native birdsong from our mobile phones, in a procession of grief and hope, over 20,000 feet pounding the pavement for wildlife, for what we have lost, for what we must do. When we reached Whitehall there were more speeches from conservationists, and more photos. The crowd was immense, stretching as far as the eye could see.

Outside Number 10, water pouring from our coats and hair, we handed over 'A People's Manifesto For Wildlife', written by Chris and many others, full of ideas for a wilder future. It was another leg of a journey that started when I was a young child – conservation has always been a topic discussed around our dinner table, on our walks, at bedtime. All the time. It is part of the fabric of my being.

At some point there was another shift of location, and I found myself sitting somewhere in Whitehall inside a large and bustling space, with just five other young campaigners, Chris and the Prime Minister's special adviser on the environment. We'd already waited an age for access to the meeting room we'd been assigned, but even after being rushed through security, we were told the room was no longer available. So we sat around tables in an open, public atrium. The noise was as clamorous as the outdoor din, which made me feel really out of sync with myself. I had to focus intently. This was my chance to speak out, to be heard by a government official. So I had to physically pull my body and brain into shape, pocket the anxiety, suppress myself until I could release it all later. I had to do it, I was determined. Otherwise sitting there in wet clothes would've all been for nothing.

The advisor seemed nice enough, but as we talked it became very apparent that politically, even though we both loved birds and nature, we came from very different places. But I wasn't deterred and seized my chance, pouring my words out on the lack of ecological education, the need for urgency in Government, the need for a complete shift in society, the need for radical change, for bravery and boldness. They weren't just my words. They are the feelings of so many of us, young and old. Those of us that care. We feel it, every hour of every day. It's heart-wrenching and exhausting, but it's vital to keep pushing on, doing heartfelt things.

As I write, the warmth is starting to seep in through my damp skin. We were part of something big. The whole day was a bit like moving on the Underground, too fast to really comprehend. But I know that I can help. We all can. Taking part is important. I feel that now. Whether or not our ideas and pleas are thrown to the wind, we still have to keep asking for change.

I take a hagstone out of the rucksack and feel relief that it's still there – the writer Robert Macfarlane gave it to me, along with a book by John Steinbeck, damp from the downpours. A gift from one generation to another. An established writer to a novice. I turn over the stone in my hand and feel the weight of its smooth weathering against my skin – when I hold it a certain way, I can see straight through it, a tunnel gouged by time.

Watching me in the hotel, Mum tells me they are also called 'Odin stones' and that they have protective powers. She says that if I look through the hole I might see a fairy or two. I laugh and put the stone on the bedside table, a companion as I write.

Wednesday, 26 September

Even though socially I'm doing much better in my new school, the educational and institutional blueprint seems the same. Sometimes, sitting in class, I feel so lethargic. I'm almost comatose. The rooms are always so stuffy, the heady perfume of teenagers. It feels like Miss Trelawney's room in *Harry Potter*, sucking the life out of me. I don't want to pay attention. I mean I do, I do want to pay attention, but my body is fighting with my brain. My eyelids get heavy, my body slips in the chair. Tedium. The teachers sound like they're sometimes whispering underwater, and I'm drowning in boredom. I'm shutting down, wandering through a trance until I come to and feel lost. What am I supposed to be learning? Thank goodness for textbooks and handouts.

My ideal classroom would have no bright colours and lots of natural light. It would have a single line of symmetrical windows, six feet off the ground, looking out to sky and birds. The space itself would be cosy, and the desks would be arranged in a horseshoe, not a circle. I'd sit in the middle, at the bottom of the curve, so I could place everyone but not have to look straight at them. There would be nobody behind me – I need to know what's happening all around. On the walls, there should be lots of inspirational quotes or cool facts. My history classroom is actually pretty close to this kind of perfect, and I learn so brilliantly in this space. I come alive. I interact. I'm fizzing with excitement. It also helps that the teacher is one of my favourites.

Science labs should be havens of curiosity and excitement. Imagine that you've wanted to be a scientist all of your young life, and you find out that when you're in secondary school you'll be taught science in a laboratory. Even the word holds such promise. You imagine a room with walls of chemicals, neatly labelled and displayed. Specimen jars. Interesting pieces of equipment all within easy sight and availability. A room of possibility, invention and wonder. But no, I've been let down by science labs. All the chemicals are in a separate store, under lock and key. All the equipment is in cluttered unnamed cupboards, and there are no curiosities, except for our physics lab where all manner of interesting objects are strewn around the benches. This is clutter I can deal with, organised chaos.

Friday, 28 September

This afternoon we were going through old photos – there are countless ones of me holding slugs before I wore glasses, all cross-eyed, well before the failed operation that was supposed to fix the severe squint I had in both eyes. It did work on one eye, which I suppose is better than nothing. My glasses cover up the remaining squint, at least I think they do, because no one has ever teased me specifically about that – there were so many 'faults' to choose from.

I've gone a whole month of school term without being bullied. The new reality is still sinking in. It probably sounds ridiculous to keep mentioning it, but this really is huge. To not carry around the fear. Usually, the desolation becomes a physical presence, a bulk.

In the forest, the greens are still the dominant colour but are starting to fade. The leaves of the beech trees are more golden every day, brittle on the branches. The gulls, rooks and jackdaws are becoming louder as the world around them recedes. The week has been a busy one outside of school. I sent the 'Peoples' Manifesto for Wildlife' to my MP and organised a meeting with him to discuss how we can do things on a local level. Then a couple of nights ago, we went to the Ulster Museum in Belfast for the Dippy on Tour exhibition – dinosaurs were one of my first passions.

The exhibition was full of natural history exhibits too, the journey from the Mesozoic onwards. Weirdly, there was a photograph of me in one of the displays. Apparently, I was an 'expert explorer'. I had completely forgotten the words I had written for the museum a few months ago, now featured beside two *actual* experts: the prolific naturalist Roy Anderson and Donna Rainey, a wildflower and pollinator hero of mine (we met on Twitter).

I love how worlds collide like this: social media is many bad things, a source of anxiety, stress and hate. But it still does bring people together and merge things that we hold dear. For me, it has been a blessing. Because I've not been able to hold conversations dexterously in the 'real' world, platforms like Twitter have enabled me to be myself, allowed my heart and brain to speak with a clarity that would otherwise be impossible. For that I'm grateful. And so here we all are in the museum: Roy with his moth net, Donna with a magnifying glass, Dara with his binoculars.

Sunday, 30 September

Silver-streaked clouds, intense cold sunlight. The beach is invigorating today. I haven't stretched my legs properly in a few days, and the comfort of walking unloads a little more weight. With every passing day, a little more joy sneaks in – is there a peak, a maximum amount of joy that we're allowed to feel? In the past, noticings or moments like this have been overshadowed, if not immediately, then not long afterwards.

Unburdened, I breathe in the salty air. The common terns are still here, readying for the journey to the southern hemisphere – Africa, Asia and South America, a round journey of over 20,000 miles. Truly epic. I watch them hover and dive. Cackling. Silver feathers glittering and dazzling, red bill piercing the surface. One tern catches a small fish that I can't identify with my rubbish binoculars, then flies off my radar as four others repeat the motion.

I lie back on the bottom of a dune bank and feel the light and the wind and the cold on my face. I feel something in the space around me change. I sit up and turn. Not ten feet away, a kestrel bursts over the top of the sand dunes. I hold it in my gaze where it stays for at least a minute, hovering. I send it a wave of admiration and it replies by holding for a few moments longer, before sweeping elegantly behind the marram grass. I bound upwards with bent body and silent footsteps, but it's gone. I fall back onto the sand, breathless and giddy. A good day. A very good day.

Saturday, 6 October

'I'm so glad I live in a world where there are Octobers.' Every year, my mum shares the same quote from *Anne of Green Gables*, and of course it's true. Outside, the world is a thousand shades of gold, glistening. Anne of the book wants to take in the maple leaves and decorate her bedroom with them, but Marilla Cuthbert (who has adopted Anne) calls them 'messy things'. There it is: these attitudes to nature aren't new. I wonder when it all started and why. Was it when we brought the wild indoors? I think we're all better off bringing nature inside and taking ourselves out, and why shouldn't we wrap fallen leaves around ourselves, bring them close, cover our beds when we sleep and dream.

Multiple vases of gathered-up leaves adorn our house, every autumn. The ivy in our back garden is flowering profusely and bees are still gathering to feed, even though the temperature is dropping. I sit on the hammock most days after school now, wrapped in a blanket and watching all the ivy-life before homework starts. So many people think that because it grows around trees, ivy somehow strangles them, inhibits their growth. Time after time, I see trees stripped of these leafy garlands, which are such a good food source and place of insulation for birds and insects, especially at this time of year.

I've noticed a few small holes appearing in the ivy from the comings and goings of our bird neighbours, now established (hopefully) in our garden. I've also counted at least five different species of hoverfly so far, the most abundant being the Eristalis species: the marmalade and sun fly variety. Hoverflies are fascinating to watch, though notoriously difficult to identify, and I always need help, with the exception of a few species.

I really am floating through the days at the moment, inside a glorious haze. Every day I wake up with energy and excitement. I have the most formidable and amazing maths teacher for a change, and for the first time I feel properly challenged. The workload is increasing, though, as I've got my physics exams for Double Award Science next month.

Friday, 12 October

A young boy of about six is playing in the forest, enjoying the fallen russet leaves crunch beneath his feet. A breeze is blowing gently, and while he rummages he finds a conker.

The boy pushes it from its spiked casing, holds it up, and the conker shines. A tiny globe of red-tinted light. The boy's mum notices, glances up from her phone, and now she's charging in and snatching the conker. 'Dirty,' she proclaims and hurls it away.

The boy is crestfallen. A light goes out.

As I watch, anger surges inside. I think about all these tiny wrongdoings, everywhere in every season, the tiniest crimes. The things grown-ups do without thinking. The messages they send angrily into the world. The consequences ricochet through time, morph, grow, shapeshift. What's so wrong with a conker?

I breathe and rise from the bench where I was watching the thrushes in the trees. I go into the pile of leaves myself to start searching, and it doesn't take long to find one, round, swollen, so perfect. The mum is back on her phone, engrossed by the glow of a milk-white screen. When I hold up the conker to the light, the little boy comes over and his eyes dare to shine a little. I pass it to him.

'Put it into your pocket,' I say. 'It's called a conker. It's the seed of that horse chestnut tree.'

In the nick of time, the boy puts the conker in his coat pocket as his mum calls over that it's time to go. I hope it gets to stay with him, if not in his pocket then in his memory. I honestly cannot comprehend where this comes from, this fear, this disconnect. Such a beautiful world, of which we are a part, is so disregarded. I think back to the meetings I've had with local politicians, their empty words and praise. I don't want praise anymore, I want action.

There's a girl on Twitter called Greta Thunberg (we've been following each other for a while now), who's been leaving school to sit in front of the Swedish parliament to strike for action on climate. She's a bit older than me but has been getting huge amounts of media attention and coverage. It's amazing, energising and exciting. It feels brilliant but frightening. I've always thought my education was my only hope of making a concerted difference to the future, my future and the planet's. My parents are not connected or rich or clued-up, and I feel so disconnected from other ways in which I can make changes, beyond what I'm doing already. Perhaps this is not enough. Maybe there's another way. A different way.

Saturday, 13 October

The sky darkens slightly and shimmers as black shadows speed towards treetop homes, a cackling of jackdaws and rooks, a coven swirling and rising and resting. Almost playfully, they are chattering on the branches one moment then surging skywards the next. I can see a new black cloud arriving, and the trees are quivering with wingbeat wind. More jackdaws mostly, with a few starlings on the fringe. The noise is blissful and deafening. Such abundance. Such life. But is this what abundance looks like? When everything was in better balance than it is today? Imagine seeing curlew or corncrake everyday, bitterns booming from the callows. Just think of cranes on Irish soil – they were a popular pet here on the island of Ireland in the Middle Ages, before they became extinct in the 1500s. Bitterns went later, in the mid-1800s when the wetlands were drained for agriculture, and then the curlew and corncrake followed. Will I ever get to experience abundance? Are we wrong to assume that our ancestors had a stronger connection to nature? They were more reliant on the fields, that's for sure. There were no supermarkets. But if we were so connected in the past, what went wrong? Why did our ancestors let this happen? Was it the supermarkets? The massive corporations? The vested interests and hidden agendas? I feel the need to be brave but am unsure how I can be. The world is so confusing most of the time. The noise, the images, the instructions. Orders, demands. All clamouring, always clamouring. Shouting above it all seems impossible. Should we all be content with changing a little corner of our world? Showing one kid a conker isn't going to change economics or the fossil-fuel industry or the other abuses of the planet's resources. This churning in me, it's got to go somewhere.

Saturday, 20 October

The leaves are falling. The days are getting much cooler, but the light is amber all day long. Mum and Dad have been trying to find new areas for us to explore, and today we headed to the forest beside Dundrum Castle. The castle remnants are still impressive, even after seven hundred years. It became a fortress and lookout for Sir John de Courcy in the thirteenth century, after he invaded Ulster and dethroned the family that I'm apparently descended from. The views out over Dundrum Bay are spectacular this afternoon, with woods surrounding both sides – beech, sycamore, ash, a few oak and wych elm.

We walk down a series of steep steps, covered in autumn colours. Other than the quick shower this morning, there hasn't been much rain lately – the leaves crackle underfoot and forewarn the surrounding creatures that humans are on the way. The smell is intoxicating. The fragmenting, the mouldering. There are some leaves still green on the beech and oak, holding on. Red valerian and comfrey bells carry droplets of moisture from the morning. I remember brewing our own comfrey for fertiliser once: I stuffed the leaves into an old pot in the garden while Bláthnaid made her own potion of young nettles and anything else she could find. And there they stayed for two years, hidden under the weight of a cypress tree. Mum, who rediscovered our brews, wretched and gagged. Bláthnaid still has jars lined up along the wall beside our back door, filled with her potions, some with white scum forming on top. Mum and Dad just let her be, knowing that there's a deeper force working up these spells. They're not experts but they were kids too, and we all know what it's like to have our feelings squashed by parents or teachers or other kids.

A quick flurry of wind unlatches leaves from a beech tree. The leaves fall and gather at our feet, as if they want us to notice this last breath of beauty and loss. We open our hands to catch some, so we can make wishes and collect enough memories to keep us warm through the winter. We sit for a while, amongst the arboreal spires, in speckled light and in silence. The sudden keening of a buzzard has us all jumping up and swaying to see where it's going. When it disappears behind the trees, instead of the usual searching, I just close my eyes and listen to the sounds, from sky to tree to ear to heart, and feel the cold in my hands.

When I open my eyes the others are climbing a bank, exploring through the bramble. I don't follow. Instead, I stray with my own thoughts and spot some bracket fungus growing on the old stump of a beech tree, a shelf fruiting outwards. I go over for a closer look at the conks, the wavy lines of colour, so symmetrical, going from brown to red to green towards the stump. Envoys of decay, polypores, nutrient oscillators of the forest. I look to the other side of me and discover a ladybird, shining brightly against *Xanthoria* lichen, like sun bursting from a branch. Its stillness begs not to be touched and I marvel from a short distance at the contrasting colours, the bright yellowy orange frills of lichen and the tiny red sleeping spot in the middle. I look up and squint to see a roughly made shape of what could be a buzzard nest up in the trees – perhaps the lichen has grown here because of bird spray.

The bird has stopped its squalling now and the forest is mostly silent again, save for a lone robin still singing. Always singing. I can see that Dad has found some cep mushrooms: supper for later. I take some photos of the bracket fungus and we head off, back to the castle grounds where we play at being knights and kings and queens, because I'm still a kid and need a battle to get my energy out. The biggest battle is loving and protecting the natural world. For now, I pour out my battle cries and mock fight with Lorcan.

Saturday, 27 October

Most places in the Mournes seem to be crowded on the weekend, but there were only a few people walking today. It was unusually warm, with almost crystal-clear skies. A single cumulus funnelled from the peak of Ott Mountain right down to the valley below. I felt cradled up there because the surrounding mountains were so close together. The walk up was easy enough but we needed a thrust to get to the top, and now that I'm back home writing the day down, I can still sense particles passing through me, waves of sound and mountain light. My hand touches moss, leaves my imprint. It's as if I am back there still, with the small mass of the experience on my skin.

At the Shimna River, which flows out of the Ott, we spotted what could only have been an otter, bounding across the field. The air was so clear and hushed that I had to lie back, close my eyes and feel the warmth of the sun. Ravens, three of them circling. Three goddesses. I feel transformed as I write myself back to the mountain, and every time I feel the vitality and beauty of nature. There is a strength growing in me, steady and knowing. I opened my eyes just in time to pick out a bird in my binoculars as it funnelled then blurred over towards Meelbeg Mountain: a peregrine, surely, its wings tucked as it dived down and out of sight. The camera clicks in my memory and there it will stay, like all these moments. Catalogued. Picture perfect. Loved with everything that is not precise or relatable, just the feeling left over every time.

Wednesday, 31 October

It's mid-term break. I've survived the first part of the academic year. Actually, I'm thriving. Perhaps that's why my feet land purposefully on the path. We're all out for a walk across the road to the forest. We've decided to do the Slievenaslat walk, the highest path there is in the park (885 feet above sea level) and it promises to be rousing and a little strenuous. Just what we need.

It's Samhain for us today, not Hallowe'en. The day when we celebrate the Celtic New Year – we celebrate Lorcan's birthday on the 'other' New Year's Eve. The early afternoon light, although not sparkling, isn't grey either. Rosie is staying indoors, just in case there are fireworks later. She doesn't react to them much when she's safely inside, but outside her body tenses, jaw starts shaking, and she roots herself even more than usual. So she's tucked up in her purple dog bed, with the curtains closed, while we stride.

In previous years, we've celebrated Samhain by camping outdoors under constellations that were usually hidden by wind and rain. There were also crazy parties with our neighbours, especially one we had a few years ago, the year before we moved to Fermanagh, which went particularly haywire and ended up with a visit to casualty for me. (I still have a scar on my gum.) I've never understood how to play games with the other kids. The rules of play are a mystery to me. And they certainly don't understand my rules, which are usually convoluted and complicated. Either

I'll underreact or overreact; I'll stand vacuously staring or get madly excitable. There is no middle ground.

The Slievenaslat route begins in a mixed wood of broadleaves and plantation conifers, but a monoculture of Sitka quickly crowds in, and there are small absences in the understorey because of the lack of light. One particular beech has thrived, though, and has a girth of three hugs (I don't know if beech hugs are measured like oak hugs, so it could be one hundred and fifty years old, or younger, as beech do grow much faster). Moss is growing up from the base of the trunk and the bark has been stripped in places. Up ahead, I see strange oak leaves on the path, more serrated and narrower than native sessile oaks. These are from the Turkey oak, an ornamental tree introduced to the gardens and parklands of Ireland from the eighteenth century.

Jays are cackling all around us. Do they plant Turkey oak as well as the thousands of sessile acorns they plant every autumn? There are Turkey oak saplings growing everywhere! Their acorns have more tannin in them than sessile oaks, which means that insects and herbivores don't like eating them; this might explain why there are more Turkey oak saplings than sessile. There are lots of deer around here, and usually they nibble at the saplings, which can be a real problem for natural regeneration of woods. I wonder if deer like eating the Turkey oak saplings?

The gravel path ahead is covered in leathery beech leaves, and they squelch or crunch depending on where the water is running off the Sitka bank. The leaf litter suddenly changes to sweet chestnut. I touch the furrowed bark of one tree, my fingers settling in the grooves. We pause by a pond, still as glass until gentle ripples are set off by fish below the surface. The water is earth-brown and surrounded by widely spaced conifers and some goat willows with low-lying, outstretched branches almost touching the water. Suddenly, there are cones raining down. We stop to look up, spotting an auburn shape tousling the branches of a Sitka tree. We crane our heads until the cold seeps in. The movement stops and the red squirrel just seems to vanish into thin air.

We move on, delighted and smiling. The gravel path stops as we cross onto more earthy and rocky terrain, over roots and rock. It's a little wilder too, with willow and hazel gorse, and bilberry showing its samphire stems (we must come back in late summer to harvest the fruits). The light becomes dazzling as we move to higher ground, ascending quickly to magnificent views of the Mournes and Dundrum bay. We can still see an ocean of fertilised fields, the luminous greens contrasting with the rugged mountain, forest and the patchwork of birch leaves in Dutch and orpiment orange. (I've been reading *Werner's Nomenclature of Colours* by P. Syme, and these two oranges were used to describe the seed pod of a spindle tree and the belly of a warty newt. These descriptions make my heart soar!)

We sit at the top of Slievenaslat, a family alone with a waning sun. Alone but for the hazel, showing us who it really is – I think it's my favourite thing about late autumn, the way it reveals the structure of trees. Angled maps, splayed vulnerability. This is what they really look like, leafless, down to bare branches.

Starlings are congregating in the distance, and their presence seems to pull us spiralling back down the other side of the mountain, appreciating each other and everything around us. I find some chewed-up pine cones as we reach the bottom of the path, red-squirrel leftovers. All at once the conifers disappear and a startling copse of birch trees emerges, a thick carpet of copper underfoot. We explore around the trunks for a little bit, touching smooth chocolate bark. Kicking up the leaves.

As usual, we've spent so long outside that we need to rush to the local pub for a late dinner, then home where we relax in the sitting room surrounded by flickering candles lit for those who have left this world. The Celtic New Year is an opening up to the dark, lit by fires, warmed by the awakening of senses, and hopefully some space to think with the stark branches of winter. Dad plays guitar. We sing and tell stories. We celebrate in our own way before Bláthnaid goes out trick-or-treating, and the glowing pumpkins tell others to knock at our door.

Tuesday, 13 November

The morning starts in a hotel by the river, with cormorants sitting on blackened trees and herons stalking with coots and moorhens. Marauding black-headed gulls make everything come alive. The day then blurs, fast-forwards to Kew Gardens, London, where I'm inside a beautiful conservatory, feeling excited but also miffed because I want to explore the plant collections and discover the birds (anything but the ring-necked parakeets). Instead, I have to fulfil my duties as an ambassador. I have to shake hands. Nod, smile. Be polite. I'm handed an award, signed by the Prime Minister, and astonishingly I'm actually enjoying it because I feel stronger and start thinking how my activism is becoming something. I pose for photos, smile. Click. Click click again. I'm about to make the speech. My body becomes rigid from the effort. I suddenly find humans really hard to interpret. I can hear their voices and know what they're saying, but so much of my energy is spent translating the noises into meaning. I start wondering if I'm traumatised by bullying. Is this why I always feel that something is about to go terribly wrong?

Today, like other days, the doubts are all unfounded, and everything passes without mishap into something glorious, pulsing with happiness. It's all exhausting of course, it always is. And now I'm finally on the plane home, I can feel the oncoming wave and there's no choice but to accept it, succumb with glazed eyes and a hurtling heart. But before I crash out, I need to write it all down.

Back to the beginning: Mum and I arrived last night. I had been invited to London with lots of other young people to become an ambassador for an organisation that supports young people and their passions, and encourages them into social action, particularly in their local community. From what I'd read, it all sounded like a good idea, which is why Mum and I agreed to be part of the 'Year of Green Action' launch at Kew.

This year I've received countless emails with requests like this, to promote campaigns, highlight an issue or write an article on this or that or my experiences that relate to another project or campaign. It's becoming a full-time job for Mum. She's had to deal with it all for me – and I know that she keeps some of it from my eyes and ears, because she knows I'll get upset. I might be a kid but I'm not gullible, and while I do want to support some amazing campaigns and people, I sometimes feel like people are using me, or the idea of who I am. Borrowing Dara to suit some endgame. I'm not a pawn, though. I prefer to think of myself as a rook: an outlier, on the margins, looking in.

My need to be independent, to keep away from people and crowds, has held me back. But it might have saved me too. I'm a young punk at heart, so the idea of getting too attached to any organisation runs against who I am. Lately, the urge to shout louder for wildlife has been growing inside me, starting to eclipse the feeling that I'm too young or too ineffectual. I've begun to feel like it could be the right time.

Still, being so engrossed and passionate and saying I want to 'save nature' is too hazy, and I still need to figure out what I, Dara McAnulty, can do to effectively make a difference.

This campaign launching at Kew sounds different. Their pledge focuses on how they can help young people, rather than the other way around. When we arrived at the hotel to have pizza with the other ambassadors, I was terrified. I wanted to run, which made the panic set in. And when this happens I overcompensate. Words just tumble out. I say too much, too fast. My chest tightens, palpitates, then words rattle at the same rate as my heart beat. From the outside, I might sound a little overeager or eloquent because facts, stories, anecdotes keep pouring out. But so does the sweat, dripping from my head into my shoes as I melt into a puddle of chaos.

I said the wrong things, I ate too much. My inner-editor couldn't work fast enough to correct the thoughts, could not steer me away from the car crash at the restaurant. Mum could see my clenched fists on the table, my shifting feet, and she knows what it means when my jaw starts to jar and my breathing quickens. She knows how to intervene in these silent wars – sometimes, all it takes is a glance or a look or a squeeze of the hand. Unbeknownst to everyone else at the table, chatting, enjoying their pizza, the raging suddenly stopped.

Escaping back to our hotel room was a massive relief. With an aching head and heart, I locked myself into the bathroom and sat, head between my legs, trying to release the pressure. I adopted a sort of prayer-squat, which is comfortable and helps me breathe. I discovered this technique at school once, accidentally, after someone had pushed me around in the playground. They couldn't help but take it to another level when I wouldn't 'play ball', and their insults came right up and into my face, louder and louder because I ignored them and turned and walked, knowing they wouldn't come after me, as our teacher was coming out of the mobile classroom. Safely away, I found an empty store cupboard and squatted out of sight. I started to breathe, deeply, and the images and words seemed to slip away a little, allowing my body to relax, the pain to fade a little. It wasn't a cure, not then, and it isn't a cure now. But it does allow me time to gather the pieces and go back onto the battlefield and try again.

Crouched in the hotel bathroom, as I was starting to feel better, that's when the words started to form in my head. *Glowing pathway, beckons . . . I walk amongst it and the radiance travels in and around . . . The growing up and growing apart . . . Suddenly, a blackbird's furrowed song*. I was supposed to be making a speech at Kew, about young people, about nature, and although I had written something already, these new words were starting to make sense. *Pathways. Ancestors. Aching. Me in the midst of it all. Healing*. I felt better. Getting these words out calmed my brain. I wasn't sure if they were the words anybody wanted, but they were my words, so they would have to do.

Wednesday, 17 November

We meet outside Dublin's Dead Zoo. There are rows and cases of dead and extinct animals, trophies shot for game. Collected. Amassed. Hoarded. Glassy-eyed. Lifeless. Museums of natural history normally fascinate me, but here I feel sick and bitter. There are lots of people, a sea of faces with placards and banners and drums. There are cheers, chants, waves of solidarity pulsing. Many speak before me: politicians, lawyers, academics, a fellow young activist called Flossie (she's really cool). This surge of humanity, this coming together, is an Extinction Rebellion.

I might love punk music and hate conformity and being boxed in, but I never saw myself as a rebel. But maybe I am, and as I stand on a wooden box, the organiser, called Carolyn, holds a microphone so I can read my speech. I feel emboldened, outspoken. I feel like it's the first time I've actually said out loud all the many things I'm angry about. It feels energised and raw as I look above the people, raising my voice, declaring out loud, my anger rising.

These are the threats we are facing. These are the crises that the most vulnerable in the global south are already facing. Yet those in power do nothing. Those in big business just carry on making obscene amounts of money. We are governed by materialism. Flocks of curlew and lapwing were commonplace when the destroyers were children, like me. But unlike me, they do not see the world as I do now. Depleted. They couldn't possibly know. Now, however, they are in denial. If they weren't, how could they carry on? The fields are falling silent and empty and although I love corvids, I want to see diversity. A healthy and balanced ecosystem. Even my beloved whooper swans are not as bountiful. I try to imagine the noise, the music, the orchestrational clamour of the song. I can't because it's not there. I ache for it. The world is still hurtling too fast. My generation will experience the worst of it: rising sea levels, oceans with more plastic than marine life, oceans starved of oxygen because the phytoplankton cannot survive the acidity of the warming water. The loss of wildlife crashing to extinction at a rate never seen in human history. Soil, where all living land life springs from, is so toxic from pesticides that insects can't survive.

My brain feels out of control. This anger was seething inside as I travelled down to Dublin. It is still seething now as I speak with the first Irish gathering of Extinction Rebellion, and I'm still nervous because it might get rough or the police might come, as they have done during the London protests. When I reach the end of my speech, I step away from the microphone to breathe. The heaviness of it all. It is wonderful, standing in the street, clamouring. What will it do though? My head hurts. I feel like the child I am, powerless and inept. Yet, I shouldn't be feeling this. This weight on my chest has been unfairly dumped. Anger seethes again, which is never a good thing. Or maybe it is.

Tuesday, 20 November

I'd struggled to concentrate all day. Things have been going so well. I was really starting to enjoy my life here at school, so why was I putting my head above the parapet? Foolish. I couldn't help the urge, though. One of the history teachers had heard about my 'work' for nature and planted an idea, left it hanging in the air. It wouldn't go away so I convinced myself to try again.

I'd tried and failed so many times at so many other schools. No one ever turned up, except the odd well-meaning teacher whose interest would eventually wane, 'It's not really my thing'. It was after one such day of deflation, while I waited for an early lift home, that some kids started their taunting and baiting. They pushed and they shoved and my face was in gravel, a sulphurous bloom in my mouth. Quickly cleaned and easily explained to Mum – I bit my lip, I bumped into something, quirkily I missed a stair. And now I was about to try again. I have to start turning the anger into something.

When school ended, Lorcan and I made our way to the designated classroom. I can't remember what happened first, but I was aware of myself standing to speak and could hear my voice boom in my ears. I stood with the kids from different year groups, some younger, some older, fifteen in total (seventeen if you include Lorcan and me). They listened to me talk about why nature has become so important to me, how I store even the tiniest noticing so I can retrieve it on demand to help me navigate everyday life, and why, because of this, I want to stand up for wildlife, shout loudly about the wondrous things I've seen and learnt, all the magic that we can see if only we stop and look. Then I stopped and stared and started to breathe, and said we should go outside, which is what we did, in the waning light after school.

They followed me out of the car park, across the damp sports fields, out of school and into the trees and the presence of Slieve Donard, where I showed them lichen on bark and explained how it was an indicator of clean air, and asked them if they felt lucky having a forest on our doorstep. And these mountains guarding us. The sea before us. Habitats of wild importance. When we found some fungi I wanted to tell them how amazing it was for all living things, but the buzzing in my ears had started and was getting louder.

My heart raced. I could actually feel my brain disconnect as I tried too hard to process the questions they started asking. How should I read the way they're looking at me? Do they respect and enjoy the answers I give? The smells of the evening air and the rustling in the trees were becoming thunderous. The effort it took to stay with them, to stay focussed, was gargantuan. But it was worth it. None of the fifteen kids sneered. They didn't heckle. They looked at me, listening. They asked more questions, and before we called it a day and walked our separate ways, we were making plans and talking about when to meet next, calling ourselves an 'eco group', and figuring out what the aims should be. As everyone left, I could see my breath in the cold night air and felt the shape of a glow around me. The herring gulls and jackdaws had all roosted, the rooks were in the trees above. The oystercatchers piped their last notes to the dark.

Saturday, 24 November

Every time Bláthnaid goes to ballet, after we've dropped her off, we amble between the sea defences on Newcastle Beach. Even looking out to the wildness of water, as the wind ricochets off boulders and pushes me out to sea, you can feel them on your back: the defences here are as unnatural as the amusement arcades and the water park. There's so much beauty around it, though. The mountains behind and the sea in front. It's a shame about the strip between, the human bit, for all the tourists and those who make a living from tourists. It has a lovely library, though, because I can still see Slieve Donard from the shelves in the children's section.

The day is overcast. The skies look ink-dipped, steely. I leave the promenade for the beach and head towards the waves. There are pebbles but no sand now – the boulders have made sure of that, and longshore drift has rippled the sand up to glorious Murlough Beach, which I can see up ahead. They're talking about regularly importing sand, which is a ludicrous notion if you think about how the coastal sea defence and promenade have changed the dynamics of shifting sands. The whole lot will just be continually transported by the waves. We can't always control nature, and here it's almost impossible.

I sit on one of the groyne planks, blackened and splintered but an okay resting place. I see movement at the shoreline, a whirring, an almost mechanical spluttering. I reach for my binoculars and see them: sanderlings, about thirty, moving erratically yet with wonderful purpose. Blurred black legs. A flash of beak prodding the sand. Sand ploughman. They whirl with the waves, never stopping. Scurrying. Rushing. Every movement too fast for me to focus on. Dazzlers of the shore.

Sanderling plumage is snow-white and pewter-back, the

crown darted with linear black-among-white. They come to winter in Ireland from the high Arctic, travelling non-stop for over 3,000 miles. Their movements are completely hypnotic, especially as I focus in on one bird and observe how it moves relentlessly at speed between the waves and shoreline, sandpeckering as it goes, and repeating it all again as the waves recede, over and over, over and over. What tenacity. I'm not sure how productive it all is, as they never stop for a second and must spend so much energy making each tack from wave to shoreline. I think of the waddling oystercatcher in comparison, constantly taking little breaks as if to enjoy the scenery or have a quick ponder about life. Of course, I know this is silly. Every species has adapted according to their environment, but I find all the differences remarkable and thrilling to watch.

The magic is broken by a black spaniel careering across the stones, off lead, acting as though it had been released from prison. Shattering the gathering of sanderlings, sending them into flight, off to find an undisturbed place. I hope they find somewhere.

Sunday, 25 November

Darkness is coming, light is becoming precious. By stealing the day, night brings with it an urgency. It starts to snatch away the songs of the garden, but also shows us places that have been hidden by the abundance of summer. I can explore these new places, hide among them as the light fades. There is still music to be heard in the fields, though, so we decide to travel out to the River Quoile in Downpatrick and listen.

There's a path leading from the car park, where the black-headed gulls flock beseechingly, whirling like dervishes, screeching, the movements tossing mallard feathers up from the tarmac. It's unsettling but I can't stop watching as they keep landing on the bins and search desperately under picnic tables. It all turns into a frenzy when a car pulls up and the lady driving emerges with a bucket full of bread. I feel each gull movement like a dart, feeding because their lives depend on it. A gull food bank. Mum steers me towards the path, away from it all, but my heart is still hammering as we walk and the ruckus is out of earshot.

I sit on a bench by the river to catch my breath. Mum goes with Bláthnaid to collect sticks, Lorcan sits beside me. He felt it too. The hunger of gulls. We chat about it until the high-pitched '*tsee*' distracts us, and we both look up for a glowing streak of goldcrest, and find it almost immediately flashing through the leafless alders. It lands and starts pecking at a moss-covered trunk, flitting again, hovering, foraging from branch to branch for insects and spiders.

When I turn around, Lorcan is gone – he's heard another sound nearby, the gregarious whistlings of long-tailed tits, a gang of them. I join him and we look on as a watery sun blasts out from the clouds and pours onto us with warmth. The birds manically flit, precariously balance, rotund body and disproportionately long tails. Lorcan and I look at each other, smiling – watching the long-tailed tits, our contentedness doesn't need any outward cues. We hold it inside, a silver spun thread binding us both.

Walking on, we find Bláthnaid outstretched, dangling over the river on an alder bough. We join her on alternate branches, balancing on air, and watch a raft of tufted ducks float by. The sky turns amber, and there is a cool crackling wind that I let descend on my chest and lips and fingers. Something about one of the ducks looks different, and taking a closer look with my binoculars I see it has a golden eye with a white patch just underneath. It dives and I follow it until it rises again, a silky black head with shimmering greens. A goldeneye duck. They're so beautiful. Perhaps it's here alone for the winter because there's no female around. We rarely think of all that effort being made below the water, those webbed propellers whirring so the bird can glide with such ease and grace on the river. It's just like being autistic. On the surface, no one realises the work needed, the energy used, so you can blend in and be like everyone else.

I tell Lorcan and Bláthnaid, pass the binoculars down the line. Each of us marvelling. Mum calls over, says if we don't go soon we won't be able to get inside the bird hide before dark. We slide off our perch, reluctantly, the three of us a trinity. By the time we reach the riverside hide the light has faded. It's still a magical place – and we're alone, so it means we can each sit at a window. We open them to let the cold in, and with it comes the sound of teal among the rushes, coots barking and bullying the mallards. The movement on the far side of the river is a flock of lapwing feeding along the bordering field, their plume-crowns fanned out. An unseen interruption and they all rise, an eruption of flickering over bronze clouds. The last of the sun, casting shadows on this mini murmuration. *Peewit, peewit. Peewit, peewit.* Sprinkling and pulsing wings, they twist around together once before coming to ground again.

The sun drops and time slips everywhere, but here it stands still and I can feel the lapwings as if they were all beside me. The world moves so fast, with too little care and too much cruelty. Here, everything is still and filled with the music of wingbeats, bird calls, the odd human gasp and giggle. The day was all golden, all light, despite the darkening skies around us.

IM SURE
THE
DINOSAURS
THOUGHT
THEY HAD
TIME TOO
ACT NOW
A-LEVELS
but there's no
YOU WILL DIE OF
OLDAGE
FROM
YOUTH STRIKE 4
BELFAST
FOSSIL

WINTER

Winter darkness, the ghost as it breathes the blast of freezing wind. Snow days are magical, but what about the rest of winter? The drained days, submerged in grey and brown, a dripping watercolour. The absence of abundance reveals contours and shape in the land. Structure, spires of bareness. Welcome in the gloaming, embrace the night as it takes up more of the day. Feel the sky closer than ever, as it presses, sometimes gently, more often forcefully. The beauty of it. The fragility of the air and the tendency of darkness to overshadow all seasons. Winter, for me, is now feeling like a time of growth, of contemplation, connection with our ancestors and those that have passed. Their stories, messages, artefacts. More darkness means more quietness in the evenings, when all that can be heard is the robin's song, the rook, jackdaw, raven or hooded crow, the distant squealing of gulls. I can hear so much more between.

Rising in darkness is the hardest part for some, but I have always enjoyed it ever since those very early childhood mornings with Mum – stories smuggled under blankets, the games of chess before sunrise, whatever the season. By the time it was light it felt like we'd done so much. Often I rose alone, to trace the sounds before dawn, the ticking clock, the buzz of the oil heater coming on, the creak of the radiators as they filled with hot water. Cogs turning to set the day in motion before the sky brightened slightly and the jackdaws danced on the extension roof. Then a singing robin. The spilling out of a Lego box. The sound of wood on wood as I laid out Dad's old chess set, the brass latch falling apart, his name written inked in Gaelic script. Getting ready in the still of the dark is the best way to prepare for the day, etching before daylight, making marks, watching the curtain of time open up before the day unveils. So much more can be seen in winter, the shiver of branches as wind travels through, more perching shapes too, and so much uncovering still to come.

I vividly remember a day in December when everywhere along and around the Lagan towpath was illuminated with so much white. I remember the coat I wore, the beige duffel one, because I loved it. Blue wellies. My curls have grown long, and Lorcan is running now. Are these his first steps? What was I? Three years old? I wonder if other people can remember this far back. To me these are the brightest memories, bell-clear, crisp as our footfall that afternoon. The sun is low but bright and there is a long stretch before we meet the willow trees, arching over the river. Possibilities hang low. An island of life approaching. Instinctively, I quieten and move more slowly; I see a rippling, unsettling the reflection of branches. Smooth back, black, slinking. I point it out to Dad and we sit stock-still. Mum cuddles Lorcan and whispers in his ear so he stays still too. Shadowy shape, otter, raising its head and swimming – we see it so clearly, and there are no other people. Just stillness and otter, otter and stillness. I feel the weight of the moment, a tear slips down my cheek. I don't know why it escaped. Otters do that. And when it turns around and disappears, more life fills its absence: beak first, a blue light darting across the river, a kingfisher so quick I must have imagined it.

This is how the sobbing starts, such great sobbing.

Winter brings it out, the clearness of everything, the seeing without seeking. The same way that sound carries further. Looking up and seeing the parts of things that are always hidden. Of course, the length of winter does take its toll. It becomes overwhelming, especially when the expectancy overtakes you with wishes for spring.

After that Otter Day, the snow melted, and it seemed that every day after that was greyer. I could still see the colours that weren't really there, halcyon, the shimmering ripples. And now, as I pass into the last quarter of my fourteenth year, I still keep the memory to pull out whenever the darkness becomes too much, when the night is more of a foe than friend and it cloaks you and presses so heavily that you can barely see or breathe. Inside, that's where I store these moments, accumulated in a cabinet of noticings and happenings, brought out when I need them most, to illuminate. I must go into the world to find new things. They are always there. Always.

Saturday, 1 December

We enter the holloway and I feel the string pull me along, the one that connects us with things that no longer exist but are still real in our mind. Recently, my inner wanderings have been spiralling and the conversations I have with myself are becoming strange and unshapely, but feel profound and electric. I keep visualising time as a length of string, with a flame burning at one end that represents the present, where we can act and be most alive. The ashes are the past, the intact string is the future. The string splits every time something happens. The dead are ashes: they still exist and never leave us. I can feel the string descending, still blazing in parts, but mostly it is crisp and brown and stretched out ahead.

In the holloway, hazel overarches and I can see the exposed roots and the earth curl around me, narrowing to a vanishing point of light in the distance: a glistening full moon. My footsteps are loud in my ears, deeply trodden into winter earth. Everyone else is way ahead but I feel like the Iron Man, clanking along into another dimension, infused with sizzling energy. A sound pierces the foot-beat: a robin in morse-code trills, an SOS messenger. I shake my head to spill out the strangeness but the eeriness remains.

A breath passes through the branches, a solemn creaking that almost sings. I start to feel really uneasy, and suddenly my senses invert as I emerge from the twilight passage. Strange shapes and colours emerge. I turn to the right and stride into the daylight. The gorse is almost in full flower here. Brambles too, with yellowing leaves. Hanging from the hazel are cuddly toys, trinkets, swaying baubles, boxes which I don't open. I quicken my step and come to a green gate with a sign that reads 'Ballynoe Stone Circle'. I walk over grass tipped with sparkling frost; it crackles underfoot, deafening, and still the string seems to tug and pull me along, fire and ashes, towards the standing stones of the neolithic burial ground.

In late-evening light, the stones form an almost perfect circle, the entrance to which is not frosty, and you can see a bold outline, a pathway in. The otherworld is so present here. As I approach the enclosed mound, the solid stones look as if they've just risen, full of life, filled with the blood of shifting soil. That's the peculiarity of time. The string can split into an infinite number of possibilities. The ancient human remains buried here, disturbed by excavation, have had their cremated ashes scattered out. The string droops and truth unfolds around me. Those that left us so long ago still exist in something. In earth, trees, the robin perched on one of the inner stones, tapping out notes. The bird hops from stone to stone, stopping for a few moments to sing.

Granny believes that the dead live in robins, or that their souls do. Grandad died when I was two years old, and every time she went to visit his grave, a robin appeared and sang gustily. It felt like Grandad, she said.

I wish I could remember back, to when his days were growing shorter and he sat in his armchair. Granny tells me that I would notice when he was thirsty and fetch his water bottle, and even brought it to his lips. 'Just a small bit,' he'd say. I snuggled on his knee and was quiet for a moment. I was rarely quiet then, always talking, always moving. But in those moments I hushed.

The day he died, my dad lifted me up to kiss his head. I

wish I could remember how it felt, and how it felt before then, when he was alive. I do remember Auntie Sharon, though, laid out – in Ireland we're not afraid of death, we embrace it. The dead are 'waked'. Their body encased but without covering. It's a celebration with lots of food (constant food), tea and alcohol. There's conversation, too, of course, lots of memories, hugs, tears. All the emotions. A coffin sits and people gather around it, to sit in contemplation. Pray. Say the rosary. Bittersweet, melancholic. I remember kissing Auntie Sharon's head and Lorcan had to be lifted up, like I was with Grandad. She was forty years old. It was September, and I was seven, Lorcan was five, and Bláthnaid wasn't even born. The cancer took her as it did Grandad. I remember the cool, paper-thin feel of her skin. That she didn't look the same. She was a shadow. Her old self had gone somewhere else, away from the thick air in the upstairs bedroom. People were sitting all around, and she was in the middle.

There is a solitary tree on one of the cairns to the north side of the burial tomb, behind the mound. It's hard to say what species it is without leaves, but like the hazel in the holloway, there are offerings dangling from almost every branch. Amulets. Talismans. Hopes. Dreams. Rememberings. Some withered and frayed. Ribbons, framed paintings of flowers, baby toys, figurines. Seeing them all there, as the sun goes down, with the wind whispering through stones, I experience the same feeling I have when an otter appears or a kingfisher returns. Stretched. Everything is stretching. Body, mind, intellect, all the space around me fills with infinite possibilities.

The dead are around me now and the veil hovers. I lie down on the cold earth, close my eyes and feel the pulse of what's beneath. My eyes remain dry though – I can't remember the last time I cried. Perhaps I've had the tears beaten out of me, dried, hardened. A sinking sensation crawls inside me, my chest contracts. Mum comes and sits beside me, perhaps feeling the sensitivity of the moment, or maybe it's because, without realising, I'm humming aloud. I sit up and stare out over the fields. The darkness tries to come closer, but I pick up the robin singing in my ear and my chest feels lighter. The weight shifts. I exhale.

I move up and off with Mum, Dad, Lorcan and Bláthnaid, up through the holloway, now using a torch. We make our way out and into a different night, with cars and streetlights, and we have to content ourselves with being able to go out and in like that, creating a thin place where both worlds intertwine.

Thursday, 6 December

Walking home from school in half-light, Lorcan and I are enjoying our days, despite the sheet-grey steel of the sky. No blue now, not for days. Light can come in many ways though, and each day we stop to hear music coming from the ivy. Sparrow song, a coven of cackling. A raucous, glorious racket. The ivy is pulsing below a leafless rowan tree, festooned around the lower trunk like a celebration. Nervous flittings, in and out, pecking at the branches. This ivy is a mansion for them, harbouring an entire flock. But this sort of congregation is no longer ubiquitous. Numbers have declined by almost seventy per cent in the UK since 1970. House sparrows make their homes near human places, and finding a tree like this one just outside the church is a blessing. Their feathers are fluffed up as I listen to the phrasings and who is singing them, in what sequence: the females go first, followed by the male birds, separately from brown to silver crown, then they sing together. The droning traffic slicks through rainwater, but the cold drops of water landing near us, on us, do little to subdue our joy in watching and listening. This group is different from the one in the shrubs outside our house – a few days ago, when I walked past the ivy, I phoned Mum to see if there was activity at the house, at the same time. Happily, both ivy and shrubs were full of chatter and twittering. Mum counted twenty-five birds. I've counted forty here. These numbers make me smile.

It's amazing to think that house sparrows have an extra bone – the preglossale – in their tongues, making them perfectly adapted to eating seeds. In Greek mythology, sparrows are sacred and often associated with the goddess Aphrodite, symbolising true love and spiritual connection. In *The Iliad*, Homer writes about a serpent eating nine sparrows, eight chicks and their lamenting mother, and thus deciding how many years the war for Troy would go on. I wonder how many people look at sparrows and feel that depth of connection, or simply marvel at how lucky we are to share the same space in the ecosystem. All birds live brightly in our imagination, connecting us to the natural world, opening up all kinds of creativity. Is this connection really diminishing to the point of no return? I refuse to believe it.

As I stood there in the rain, fluffed-up sparrows in conversation amongst themselves, there was a spark. Noticing nature is the start of it all. Slowing down to listen, to watch. Taking the time, despite mountains of homework. Making a space in time to stop and stare, as the Welsh poet W. H. Davies wrote in 'Leisure':

> What is life if full of care,
> We have no time to stand and stare.
> No time to stand beneath the boughs

And stare as long as sheep and cows.
No time to see, when woods we pass,
Where squirrels hide their nuts in grass.

I don't see it as 'leisure' though. This is good work. Heart work. Taking the time to observe nature, to immerse oneself in its patterns, structures, happenings and rhythms. It's how mathematicians and scientists are nurtured. Alan Turing studied the patterns in nature: the spherical organisation of cells in an embryo, the arrangement of petals on a flower, the waves on a sand dune, spots on a leopard, the stripes on a zebra. He was looking for a mathematical formula for the development of cells in living things. He called it the Reaction Diffusion System, the transformation of pattern into stimulating reactions. The complexity of it! There's no way I can interpret his theory at the moment, but it was contemplating nature that inspired him and his ideas. Nature sparks creativity. All we have to do is start with the question, Why? The way my mind whirrs and whirls in nature, or even when 'daydreaming', is way more productive than the work I do in school.

I consolidate myself by thinking, and thinking whilst intensely watching the flight patterns of dragonflies or starlings is explosive and mindblowing. Who knows where watching sparrows will lead!

Garlands of ivy, heart-shaped leaves, laden, some still studded with flowers, many with blackberries. A thrush appears and pecks at the fruit on the ground level. A blackbird, appearing in the middle of the ivy to eat more. My school clothes are soaked through now, and I realise Lorcan is gone. He probably told me but I didn't hear him. I was too busy. This pause between the bus stop and home is so much better than the homework that awaits.

Saturday, 15 December

The rain is falling in sheets, crashing on the ground like glass. Still the birds come to the feeders, guzzling so ferociously that we have to fill them more frequently. Seeds, nuts, suet. It is an avian food bank. The rain is blurring the view, so I open the patio doors and pull a chair to the entrance, and then pull another one up so I can put my feet up. I'm simultaneously reading through my maths problems for the weekend and watching who's visiting us, marking and comparing the weekly tally. Most weekends I do this, especially if the rain is belting down.

Despite the lashings, the two coal tits come and go, first scraping their beak off on the wrought-iron handle of the bowl which houses some of the seed, then gathering some and flying off with it. The robin sneaks below, pecking at the ground mostly, but also landing on the bowl sometimes – never the feeder. The dunnock is the same, always lurking, and has never come to feed at the bowl. Amazingly, a wren lands near where I'm sitting and pecks at the line of food accidentally dropped when I tipped the bag over, distracted by the craking of a raven overhead. The wren comes so close I can see the undulations of white and brown, I can see the wind blowing through, lifting up its feathers, despite how wet they are. Its tail cocks a little, something has changed. I'm statue-still as it looks in my direction and lifts in one movement, off and around and into the Forest Flame shrub beside the door, long established before we moved here (probably bought from a garden centre).

Although not great for a wildlife garden, the Forest Flame has so many bursting-through holes that birds have made, going in and out at speed. Some of the holes are tiny, some are much bigger. What do the birds all do in there? Sometimes, it sounds like a brawl has broken out inside a Tudor inn, so many different 'accents', residents and visitors alike. A multicultural community. All different, all bird. Once, a rook flew in and the whole thing exploded with wings and screeches.

A great tit pops out and starts where the wren left off, chaffinches joining in. The pair of siskins are still here, and one flies in to feed, usually the female, with the male watching on at a distance until they both fly off. I look down to see that a pool of water has accumulated at the entrance. The rain is still pelting now, and my trouser legs are soaking. How did I not notice? My trance world yet again. Time passing in a vacuum of watching. I'm glued to the seat as three jackdaws fly in, bright-blue eyes. A magpie hops in next and they're all there, dishevelled and communally feeding, shaking their feathers, sending sparkling droplets of waterlight into the greyness. A feeling ripples through them (or they're just full of food) and they all leave at once. Their absence lets the sound of traffic into the garden, and makes everything feel hollow. Shivering, I close the door and change out of my saturated trousers.

We can create a safe space for nature in our gardens, especially during the winter months when food is scarce. Caring for nature and for ourselves can happen anywhere and everywhere: gardens filled with life, nature reserves, resting spots, feeding spaces, nourishing places. Focusing in on the activity and behaviours of wildlife in our garden is so satisfying, for the mind, for the heart. Homework doesn't feel like a chore after time spent quietly feeling rain and watching birds. There is nothing better than tending to this connection between all living things, and maybe even ensuring the survival of some species living in our back gardens and along our busy streets.

Sunday 16, December

The skies cleared today, breathing light into the continuous grey of previous, dripping days. It feels like we haven't been out for a proper walk in weeks; a claustrophobia compounded by the end-of-year tests and the ashen weather. We've all been going a little crazy. I've still been walking through the dankness: early mornings up the hill in the dark, as far as the horse trail in the forest park and back again. Brief moments of escape into the rain. The rush of wind always blows away the lethargy and apathy that builds between walls at home and in school.

It's a massive relief when news of an outing spreads through the house. We're going out! To Rostrevor, the Fairy Glen. The whole family, Rosie too, in her purple-spotted thermal coat. Granny lives close by so we're going to have dinner with her afterwards.

In my early days a new plan like this would have meant absolute brain chaos. It wouldn't have been possible. Quick transitions were pain-thrusting; processing quickly seems so natural to most people, for me it was blood-curdling. Now, with gentle teaching from Mum, which required full explanations of outings and lots of planning, I can manage spontaneity with much more ease. I don't think people realise what needs to happen behind the scenes so 'we autistics' can look like we're doing alright. Mostly though, we hold it in, so controlled, until we reach a safe space. Then release the pressure. A rushing river and an easy stroll. I think of how Virginia Woolf's characters in *Mrs Dalloway* are bound together by the act of walking outside in London; my intertwining isn't with people but with the elements, with nature, and it has become inseparable from my daily life, my own story.

I have been interacting with people, though. More so than at any other period in my life. The eco group at school has grown to over twenty students from all year groups. And then there's coding club, the Amnesty International group at lunch, as well as wanderings with my friends at break – yes, friends is plural! On the surface, life seems normal. But my thinking has started to delve more deeply. Because I'm not worrying about the day-to-day so much, I've freed up space to think, to dream, to wander inwards. It's intoxicating. Spring and summer, the seasons of sun and daylight, brought me despair. Darkness has brought comfort and healing. I don't 'socialise' in the way other kids do, meeting up after school, snap-chatting, bickering about YouTubers. I'm just not built to make small talk like that, and at the moment I'm pleased that I'm not. There is so much to distract us from ourselves and the nature around us, although it doesn't mean I don't like video games – we need to be three-dimensional human beings, don't we? Multi-layered. And our connection to nature can combine with technology. There's no need to isolate us teenagers by constantly berating our digital habits – and if you are going to, please check your own habits first. Instead, provide opportunity and space for us to explore, and give us an education system that acknowledges the natural world as our greatest teacher.

Outside, the blueness is blinding after being in the shadowiness of indoors. The riverside is busy with Sunday amblers, relieving a similar itch to be outdoors. There are so many expressions of 'Ah, isn't it great to get out after all that rain.' People are smiling, enjoying the new brightness now the veil of rain has lifted.

Lorcan and Bláthnaid scamper across the river, playing on stepping stones despite the gushing. They have perfected the leg distance needed for each hop, and there are no falters as they travel across the river and back. Watching them, Mum takes sharp breaths every time one of them almost slips, while Dad – amazingly – says nothing. The need to release energy is palpable.

I sit on the cold rocks and take my socks and shoes off, letting my feet sink under the coolness, feeling the water gush over my legs. I take out my binoculars and scan the edge. Nothing. I just enjoy the feel of whirling current and the sharpness of the water numbing my skin. After a while, it starts to tingle, just a little too much. I stand to join the others but they've stopped a little further ahead to look at something, so I stay put. Then it just hops in front of me, on one of the stepping stones in the river. A dipper. Cocked head, white throat. Bobbing. The bird dives under and I can see its shape in the water, moving, walking, suctioned onto the stones with its feet. It rises up, hops onto rock and starts preening hyperactively, bobbing frantically. A flash of sherbet and silver on the bank lures my eyes away: a grey wagtail scuttling like a boy racer up the bank. When I turn back to the river, the dipper has gone and I realise my feet are nearly blue with the cold. Scrambling on socks and boots, scampering painfully forwards, I don't tell Mum and Dad about my frozen feet. I don't want them to fuss.

The Fairy Glen is an idyllic place. The entrance beside Bridge Street is bordered on one side by ivy-clad cottages, stone walls and a grassy bank, and on the other by Kilbroney Park, with the river of the same name running like an artery, dividing village from parkland, wood from forest. Oak trees line the riverside, and the beech here still have crinkled leaves, burnished and golden, not quite ready to drop. There are still lots of people so we decide to rush to the other end of the river, turn off and up to the meadow where we start to lose the throng of Sunday strollers. We go a little further, just in case, and then stop to take a breath.

Lorcan really can't stand crowds, especially outside. He can't let his chest open and his heart beam out. He can't feel the thrill of nature, he can't speak out. All three of us stim – 'stimming' is a word used to describe the self-stimulatory behaviours of those on the autistic spectrum. Lorcan uses sounds, squeaks, grunts, purrs, whistles, groans. Bláthnaid twirls her fingers, hand-flaps and makes sucking-in noises which she calls 'fluorescent stress-taking movements'. It's not weird. It's just different. Some neurotypical people talk incessantly – so much small talk! I curl my hair, jump randomly and sometimes, embarrassingly, do a few wiggles. I control it when people are around. Lorcan is beginning to stifle it when others are around. Bláthnaid, because she's younger and less self-conscious, is an unbridled stimmer. But so what? It's who we are. It's how our happiness bubbles out, and how our anxiety seeps. It's just how we regulate our brains. You probably stim too, without realising. Ever bite your nails? Twirl your hair? Pull at your ear? Yep, thought so. Maybe we're not so different after all.

As we walk through the sleeping meadow, everything seems empty at first. Then the movement begins, flittings of colour and shape. Brown, hints of red, ash-grey. Fieldfares. Redwings. Mistle thrushes. They're all pottering about. Neck up, side-to-side looking, dip to prod the earth. It rained this morning, so worms are plentiful, but when a dog barks close by they all rise up, at least a hundred, many more than I spotted when they were on the ground. They land further up the meadow but it doesn't take long for them to start edging back towards us and the field.

Sometimes called 'winter thrushes', fieldfares and redwings come here from Scandinavia and Continental Europe, and I remember the worst winter since we were born, back in 2010. We had frozen pipes and no water. We used snow to flush the toilet. On Bláthnaid's first birthday, a friend brought us bottled water from the other side of Belfast, because ours had run out. Temperatures were down to minus ten celsius, but it felt exciting to me, and we only had a fire for warmth in the living room because, like others with split pipes, our central heating had seized.

Dad remembers a particularly horrific day, walking home from work alongside the Lagan River, when he saw frozen redwings dead on the street. More were stumbling and banging into walls, or falling onto the road, dying. It upset him terribly. There was nothing he could do. He tried, but the life inside them had gone. They had arrived here to get away from adverse conditions, to find warmth, shelter and food. Instead, they perished. I'd never known such coldness. In our tiny house, cuddled up, we cried for the redwings and all the other birds.

We stand in the meadow and watch the birds for a little while longer, very much alive and well, before Mum and Dad remind us we have to go to Granny's for dinner. How lucky we are, to always have somewhere warm and welcoming to go.

Friday, 21 December

It's very early and I'm in the forest park before school, looking for little pockets of colour and light. The rooks are waking now too, splitting the air. This morning, I don't enjoy their call – I would never *not* want to hear them, but the sound feels icy, gnawing. I zip my coat up further – bright-blue, it's the brashest thing around. The grass is damp below foot, the lake swirling with waves and almost black. I feel sunken in. I came out for comfort but strayed away to the edge of feeling safe. An eeriness settles.

I scramble back, away from these feelings, and as I do a brightness comes. I look at my watch, it's late. I really don't want to go to school today. But it's the last day, a half-day.

I drag my heels until light relief comes at breaktime, while I'm wandering behind the football pitch. It's a nice spot at school, and even better now that there's a bright-blue sky above, freezing but cloudless. I lean on the trunk of a beech tree and feel its silver bark on my back, through my jumper and blazer. Thinking about today, I realise I've forgotten about the solstice. Or maybe I hadn't. Maybe that eerie walk this morning was something to do with it. I was pulled out of my bed, up and out before everyone, drawn to the steely lake for Yule, Alban Arthan. I went for a walk in a dark wood; the Druids gathered mistletoe and burnt the yule log, gathered and dusted with flour, doused with ale, set alight by a leftover piece from last year.

When I get home from school, Mum will have collected holly and ivy. The evergreens. Our Christmas tree will be up too, taking up the whole room, pine needles everywhere. It's exciting, having a whole tree in your house. We normally have a fire burning but we don't have one in the new house. It's the first time we've not lit a fire in winter, and I hadn't realised until just now. But neither had I realised how much I embraced darkness, and from today this will start to fade away. It's a turning point. The light is coming, and at home there will be candles – and Christmas. It might be the year's darkest day, but there is always light. Darkness and light. Both needed for respite, for regeneration.

The school bell pulls me out of daydreaming. A robin chimes too, telling me about the coming of midwinter, perched at eye level on a beech branch covered in moss and lichen. The bird doesn't move when I do, it keeps going, and as I bound back to school I can still hear its trilling and wonder if anyone else hears it. I stop in my tracks, run back on a whim to hug the beech tree and thank the elders that have watched over me these last four months, the best four months of school I've ever had.

Tuesday, 25 December

Christmas day dawns with Bláthnaid's excitement filling the house. A bike! A bike! And soon she'll be outside, when it's a half-respectable hour, with all the other kids, riding around in the rain. I'm always up early, and Christmas morning is no exception. The excitement is heightened and all the usual noises are drowned out by the ripping of paper. Two presents from Santa each. Chocolate coins, stockings filled with sticker books, a pack of cards, an orange or homemade gingerbread men, Lego or Playmobil figures – I never 'played' with the figures. I would build them and line them up or arrange them in varying formations. Lorcan always played with his, though, fiercely. Brothers, both autistic, but not carbon copies.

I think Christmas mornings have always been happy. I can't remember a distressed one. I was always with my family, in my house, safe. Every year we watch *The Snowman* somewhere on the telly. Last night, Christmas Eve, we all got a small pile of new books to read over the holidays and over the last of the winter months. It's another annual ritual. I got *The Book of Dust* by Phillip Pullman and a secondhand *Cadfael*, along with some nature books and fantasy novels.

After the presents are opened, dinner is prepared early so we've got plenty of time to wrap up and go for a walk. Mum is the designated and self-appointed cook – nobody argues with that, but we all help with peeling vegetables and tidying. Well, a little bit – this Christmas Mum and Dad gave us an Xbox because we like to play open-world and strategy games together. Sometimes we get violent, but we always try to be pacifists first. Negotiate, compromise. Is this like the real world? We *can* differentiate. Most teenagers can. We play on computers but we also get bored of them, and that's when it's time to go outside. Which is exactly what we do, after we've managed to persuade Bláthnaid to leave her bike behind so we can take Rosie out.

We all head into the eddying wind, aiming for Murlough Beach. It's raining when we get there, and the sky feels as if it's pressing against our heads. We usually walk the opposite way, but today we turn right towards the beach and walk over the slick boardwalk. When we reach the dunes, Bláthnaid finds some dogfish egg cases – mermaid's purses – and a raven feather. I find a kestrel feather and think back to the bird I saw here in the autumn. I stroke its strong compacted form and put it carefully into my pocket – I've never found a kestrel feather before.

As we walk down towards the shoreline, dunes on either side, the sea fret comes from nowhere and sucks up the horizon so we can only see a single strip of waves, frothing and spewing. On the beach, the wind whips at our ankles and our faces, punches our stomachs. We run at the waves, turning just in time to miss. Lorcan and Bláthnaid have found some dead seaweed and are whacking each other with it, giggling hysterically. I leave them to it and continue back up the dunes.

The mist envelopes me as I walk, curling up from the waves, smothering me in wispy tendrils. I can taste the salt and hear the crashing, but I can't see more than a few feet in front of me. I can feel the vastness of what I can't see and huddle down to take shelter by a perfect, intact dune.

Shapes suddenly burst through, rainbow scarf and hat: Lorcan in Viking mode, charging at me. I run too and we all roar into the mist, roaring for a better world. The cry is half-rallying, half-despair. It's also for the depth of feeling we have, for this place, for each other. We hold hands and run down the dune path in a line, unbroken. We are all warriors now. We run to the waves, wind slapping our cheeks red. We stop just shy of the shoreline and hug each other. Sometimes it comes over us like that. An uncontrollable urge, like a bodhrán rhythm, with flutes and fiddles, drifting in from somewhere else, wrapping around us. With wind buffeting, we laugh and pull ourselves apart, running down the length of the beach towards Mum, Dad and Rosie.

We return to the car park feeling euphoric, breathless. There's chatter in the trees, but I have to stand very still at the fence bordering the reserve to see into the mist, making out the shapes of twite or linnet, one perched on every bare branch. Little clockwork movements, a chittering chorus. They rise up and land in the field and start busily scuffling at the ground. A song pierces the chatter, as bold as a robin's but this one is a dunnock, throat pulsing with effort, and mist all around it, lilting with phrasings, bursting through matter. I do one of my little wriggly jumps because no one is here to see it. Skipping to the car, I realise I'm ravenous.

The day spins with no fuss, no stress, no table decorations (besides crackers) or party games (other than draughts, the Sleeping Queens board game that Bláthnaid got for Christmas, and of course the Xbox that Mum starts to regret buying when Lorcan and I sing the *Skyrim* theme tune). Later, looking at the photos of us at Murlough beach on Mum's phone, I can see the wind whipping the marram grass, the dunes sculpted by aeolian erosion, and although our family is small and insignificant in the wide shot, by looking closer you can see how alive we felt.

We end the day with Mum reading *The Dark is Rising* aloud by candlelight – she's even more emphatic than usual in the way she reads. Maybe it's the red wine.

Friday, 4 January

It's late afternoon. We're following red kites as they flit from tree to tree, field to field. We watch them from a distance, perched like statues. So far we've counted seven birds, but many have moved from last year's roosting spots, and we're a little early in the season for them to settle on new roosts. Amazingly, I spot a leucistic bird, all white, standing out against the trees but blending with the sky.

We're here with our friend, Noreen, who knows every red kite secret there is to know. I met her here this time last year while doing a red-kite roost survey. I remember it vividly. It was as clear as it is today, with the evening sky blazing over the Mourne Mountains and red kites passing just a few feet over our head in lazy, dawdling flight. Their slow motion let me pick out details and markings, even feather breeze. It was breathtaking. We counted sixteen individual birds on that tungsten evening.

Red kites were the first to pull the string and lure me into the world of raptors. Aged six, I started reading everything about them, learning all I could, and planning how I could get closer to them. I wanted to understand them. I wanted to help them. Red kites were once extinct in my country, but in 2008 birds were collected from Wales and reintroduced to the Mourne Mountains after a 170 years of absence owing to persecution. Our eyes can once again see these resplendent swallow-tailed birds, and we're able to spend time watching them stream in and out of sight, diving into our imaginations.

A decade since the red kites were reintroduced, their story is a tapestry of despair, endurance and hope. There have been poisonings, there have been shootings. But a small group of dedicated people refused to give up, and now the community around here are fighting back because they've become fiercely proud of 'Our Kites'. I feel part of that community too, and going back to witness their flight is a privilege. I'll never grow tired of their wingbeat.

We watch them for a little while longer before Mum confesses that she's got an itch to see starlings. She can see small groups of them in the distance, starting to gather. It's murmuration season. She starts telling Noreen and one of the volunteers about our searches to find the famous roost around here, and that so far we'd found nothing but a few strays and stragglers. Noreen smiles and gives us the new location.

The kites are hunched on boughs, not budging, looking intent on staying put. I'm reluctant to leave, a little disappointed that this evening was not as spectacular as last year's display, but as we head to the car and drive off, I can feel that familiar tingle. I've never seen a murmuration. We've always been too early or too late or completely in the wrong place at the right time. Maybe this will be the night. Will kites lead us to starlings?

We drive on the narrow road with brambles on each side, then climb to where we can see over the fields and trees. At a sudden dip, a dark cloud is moving. Mum parks at the side of the road, we get out and listen as wingbeats shatter the country silence. Flying in and around our heads, a river wind, they rise and fall on the barn roof where cattle are feeding on silage. Spiralling, they move further on up the hill, so we chase them, running, feeling air sharp in our lungs. We stop by the twisted branches of a hawthorn hedge, and look out and above to the starling shadow funnelling and sweeping across the sky. Confluent wing beaters, shapeshifters. The magnetism, their gathering in for safety, fails as a peregrine cuts through. The starlings pull in opposite directions, splintering, slithering. Again the peregrine torpedoes, and is gone. Stunning. Mission accomplished, perhaps.

As the starlings regroup there's no way of knowing if one of them has been taken. The sky grows darker, the starlings still bloom and bellow, origami shapes against the limestone greys of the sky. When the pulsing starts to slow, Mum and I can see them start to land on cypress trees, small groups at first, then all of a sudden they're all sucked away for the night, taking with them any warmth left in the evening. A deep silence replaces them, turns the night to basalt. We drive home in the highest of spirits, the dark of the night illuminated by our smiles and chatter and 'Oh my gods'.

Sunday, 13 January

A few days ago we had a warm spell which conjured up a patch of lesser celandine, unbelievably early. I couldn't celebrate them. Not really. It was as if they were growing in the shadow of a planet that's out of sync.

This morning I feel exhausted. These days, Chemistry is the story of my evenings. Homework, revision. School is still okay but I feel a simmering inside. Is the social interaction starting to take its toll? Perhaps it's the constant flow of people asking me questions, both in the real world and on social media. It's overwhelming. My ability to process it all is slowing, and it feels as if there are many areas of blankness spreading in my memory. This worries me. I manage one thing – a speech, an article, an interview, and another comes my way, domino after domino. Things are spilling and I'm starting to step outside of my usual boundaries, but my brain is short-circuiting. Too much of everything. I need to reboot, rebuild. I even have to haul myself outside at the moment, a deadening weight in my feet. It's like dragging lead. The week stretches ahead, and it feels endless. I'm trying to find it by walking and writing. Most of my days include at least one short walk to the promenade and beach, or up to the forest park to feel the wind and find the words. Writing it all down, spilling it out, helps me make sense of the world. What started as scribbles and scratches on the page has grown into an essential shape in my days. I need to find energy from something, somewhere.

Saturday, 19 January

It's at height, among clouds and granite, where I find the energy I need, on Hen Mountain, after an exhilarating run up without stopping, facing into raven land with proper, expansive wind at my face. Looking at Dad when we reach the top, I wonder if we went too fast, so we rest a while. It's just the three of us today: Lorcan, Dad and I. (Bláthnaid wanted to stay home and play with her friends; Mum was disgruntled but had to stay back.)

The walk up to Hen and neighbouring Cock and Pigeon rock is a steep one. You rise and rise and feel the stretch in your legs, and you need a blast of energy to scale it fast. This is Lorcan's dream climb, fast and furious. He wants to be a fell runner when he's older, and watching him here I can totally see it. He really changes when his energy is released. And there are very few people here – Slieve Donard is always packed, and it's hard to lose yourself. This place, though, doesn't get the same crowds, especially in winter. Maybe Hen Mountain is becoming our new Gortmaconnell or Killykeeghan. A more grown-up playground.

We reach the top, where carved-out granite chutes lead deep into the tors' heart. Three rocky outcrops, shaped like a crown, forged from fire, sculpted and weathered by time. As my hand runs along the rock's coarse surface, it doesn't feel wet but it leaves behind an imprint of moisture. The mountain leaves its mark, water transfusing out and into me; every touch and tingle is nourishing.

Between two of the rocks – two 'bull horns' – there's a bog pond, winter-still. I put my hand in and feel the peaty cold. The feeling on my fingertips reminds me of a Seamus Heaney line from 'Death of a Naturalist': *That if I dipped my hand the spawn would clutch it.* We'll need to come back in spring to see if there are any tadpole stirrings.

Whenever I'm high on a mountain I make an agreement with myself to leave behind all human worries, problems, thoughts. They mustn't veil my experience of nature, of this place. Learning to do this took huge effort, and it doesn't always happen, but doing it lets everything sweep in. I glean every smell, sound, flutter, flicker, until it takes up all the space in my head.

When people ask me why I experience nature so intensely, the truth is that I only know I've experienced it when I'm writing it all down later. The intensity gushes out and I feel everything again. I relive moments by scratching them out on paper or typing them up. I don't need to think about it much; all the details are right there in my mind and it surprises me every time. And high up here I'm not thinking. I'm feeling, observing. The brain camera clicks away at the puffed-up cloud on Cock Mountain, the hollowed indentations in the granite that hold pools of water, the shadows of the surrounding Cock and Pigeon, all the things that catch my eye.

We jump off some of the outcrops – several have thirty-foot drops, so we sit on the edge instead, dangling our legs out, feeling no pressure underneath. It's exhilarating. As we scale and rest on one of the bigger tors, a raven lands near Lorcan. I can see every feather, iridescent purple-black in the light. I've never seen one so close before. It feels like my heart might burst, or move in the wrong direction. I steady myself, take it all in. I can hear wind brush its feathers, the hushed sound of throat plumage ruffling out, and that incredible black eye, unblinking. Lorcan (for once) is speechless. He's squeezing my hand, to control the urge to shout out. It rests with us for about a minute, a marathon minute. A mountain minute. Because time slows up here, there's no rush whatsoever. No need to hurry. I hear wingbeats overhead as another raven lifts, silk rising. I watch the two elevate and caw off together; Lorcan and I can lie back and exhale everything we'd been holding in.

The rebuilding is not yet complete, but I feel more sturdy, more relaxed. My smile is definitely broader.

Sunday, 20 January

Such a heavy, deep sleep last night. I can count on one hand the very few of these I've had in recent years. It leaves me even more refreshed and a little stronger, and this morning Mum says she needs to get out and go somewhere, a little further than usual. She suggests Castle Ward, a National Trust estate made famous by *Game of Thrones* – I've never seen the TV series, I'm a little too young, but I can imagine it all. Castle. Courtyard. Turrets. And when we arrive, Lorcan groans at the sight of the tour buses and people walking around in costume. They all want their little bit of the screen magic! A selfie for Twitter or Instagram. I hope they know that magic exists everywhere.

While we wait for a crowd from one of the coaches to go on ahead, we sit on a bench and look out to Strangford Lough. We can hear redshanks whistling and curlews lamenting, their winnowing calls blowing off the water. I watch one prod the mud with its curved beak, foraging. Remarkably, curlews and redshanks have almost bendy bills. The final section can flex upwards, independently. It's called distal rhynchokinesis, the way the upper jaw moves, and even when buried in mud or wet sand, the beak can open and grab food. These adaptations are fascinating and wholly mind-blowing.

Later in the day, once we've explored the castle and grounds, we find an earwig with her eggs nestling under a stone from an old wall. Most of the wall is covered in ivy-leaved toadflax (also known as mother of thousands). A native of southern Europe, it has naturalised in Ireland over several hundred years, and is growing here at Castle Ward with its ivy leaves and three snapdragon petals, scrambling over nooks and crannies. In amongst it all, the female earwig patrols her batch of butter-yellow eggs – they're diligent mothers, and if the nest is disturbed and the eggs dispersed, she will gather them up into a cluster again then continue to stand guard. Woodlouse are important lumberers too. They also break down decaying matter, recycling, tidying. Such essential, intricate parts of the ecosystem.

A wall is an entire world to an insect, a universe brimming with life in winter. Looking closely, noticing, brings everything to life. The tiniest creatures can be the most interesting and easiest to observe. Watching micro-dramas play out, many questions rise up to the surface. Woodlouse look like they're riding on the dodgems: it seems random but maybe it isn't. I remember a battle between centipede and earwig in my garden in Belfast. I lay on my belly and tuned in, utterly transfixed. I don't know how long it took, but the earwig stabbed the centipede on one side. I wasn't upset by the death, knowing it was nature. Balance. Order in a wall-shaped universe.

Sunday, 3 February

I'm processing it all, under a big oak in Mount Stewart, a nature reserve overlooking Strangford Lough. I can hear the brent geese honking, the sound entwining with hard wind rushing up from the shore. It's icy cold. The sky is clear, duck-egg blue. Branches look startling and crystalline, like intricate maps or dendrites outreached to the air. If only we used trees more, to guide and inform us, to teach us about community, about interconnectedness.

A buzzard soars and wheels over the field in front of the oak, while another joins in and they fly together, spiralling in courtship, touching talons and wings, rising up, falling down. The beguiling display draws out some grieving in me. Some things are so right in the world. I need to hold on to all these moments, to stop myself eroding.

February has rushed in, following days of so much *doing*. My chemistry exam is over and I'm just back from London again, for another speech and event. Exhaustion is setting in. One of the more bizarre things was meeting the Environment Minister at London Zoo. Of course, he was late. Very late. His speech was persuasive, tripping off the tongue with ease. But words are all too easy sometimes. They can shapeshift and are easily forgotten without being translated into action. The Minister made grandiose promises and plans that day, but where are they now? And he never stayed to hear my speech, or the speeches of any of the young people. He was swept up and shortly afterwards it was as though he'd never even been.

Luckily, the Galapagos tortoises who live in the Zoo gardens redeemed the day. Stroking their hard shell and feeling the smooth symmetrical outlines was a relief because the rest was just a photo opportunity for the Minister and the Zoo. For me, though, it was a chance to get close to magnificent creatures, three of them, that I'd only seen on television until then. The largest tortoises in the world. I can't bear to think of Darwin riding on one, let alone eating their meat.

Like so many events I'm asked to go to, that day in London felt like a hat-tip. Kids are invited to 'have a voice', to share their ideas, hopes, dreams, anguish, and then very little actually happens. The adults never dole out an invitation for us to sit down and plan things. We hand over our hearts, beating on a platter, for nothing. At least nothing tangible. Globally, we have lost sixty per cent of our wild species since 1970. And it's my generation that is labelled 'apathetic', 'self-indulgent', 'less focused'! Whereas the adults, who are actually in control of our access to wildlife, the boundaries between busy roads, housing developments and green spaces, carry on making decisions and spending public money in conflict with nature.

The gap is ever-widening. It feels like a ticking time bomb to extinction. Is it any wonder that almost a quarter of young people are experiencing mental health difficulties? Our world is increasingly divided between attainment, materialism and self-analysis. We're at a tipping point in the relationship we have with ourselves, with each other, and our world. A world which is so intricately connected, so interdependent, so intrinsically linked. So delicate. The power struggle between huge organisations, economics, development and the species we share our planet with is growing so out of control that it's easy to become overwhelmed, depressed and disconnected.

I battle with it all the time. Sometimes my heart beats as fast as a dragonfly wing, and my mental health has really suffered because there isn't anywhere to express these feelings of despair at the inaction. My intense connection to the natural world does ease and alleviate these debilitating emotions. When I'm immersed in nature, I am less focused on myself and more aware of the other organisms around me – trees, plants, birds and fellow mammals (if we're lucky). During these encounters we experience joy, and it is perhaps in these moments that I understand so clearly that we are all in a position to make sure that this magnificent beauty is cared for, protected. We are all custodians.

I've also found that focusing in on a local level, on my immediate surroundings, is where I can be most effective as a force for hope and change. When I started the eco group at school, I didn't know if anyone would turn up because I assumed that other young people didn't care. I was so wrong. Perhaps I was haunted by my previous efforts to set up environmental groups at a different school. I now also realise that teachers are so stretched, yet even though we still need their help and the help of other adults, we can also take action on our own. The eco group is now full to bursting with all ages, and those who have joined tell me it makes them feel good to be a part of it, to put ideas into action, to share how they are feeling, to fight back. Maybe they were just waiting for the opportunity. Maybe we all need more opportunities to take meaningful action.

In a fast-paced and competitive world, we need to feel grounded. We need to feel the earth and hear birdsong.

We need to use our senses to be in the world. Maybe, if we bang our heads against a brick wall for long enough, it will crumble and fall. And maybe the rubble can be used to rebuild something better and more beautiful, enabling our own wildness. Imagine that.

Friday, 15 February

I'd never stood so still in such a cold wind. Alone in my uniform, on a school day during school hours I was clutching with gloved hands two placards which read 'School Strike For Nature' and 'School Strike For Climate'. Not a cloud in the sky, yet the strongest of all this winter's wind was blowing, challenging gravity. Blowing me, blowing sand over the sea defence wall on Newcastle beach. Four hours, I stood. Stood up to the avaricious world. Stood up to those that take instead of give. Those who steal my hope, and steal hope from future generations who will inherit a planet so extracted, diminished, less bountiful. People stopped to ask me why. Passers-by, teachers, parents, radio stations wanting interviews. I wasn't expecting that. Instead of talking about the issues, they wanted to talk about 'me', how 'I felt'. Not the science or the facts. Not the abomination of climate change and mass extinction, or why young people around the globe have been forced to act – young people who value education profoundly but are nevertheless compelled to act against the inaction. I'm not a doomsday prophet, though. I can't be like that because I see so much beauty every day, and this is a huge privilege. I would never question anyone's grief or fear, because these are real things too. Millions are already facing an ever more precarious existence in the climate catastrophe that is manifesting. Their experiences are real, their fear is real. How will those waves crashing over the sea defences behind me be in ten years, in five years? How will everyone in this seaside place be affected? So yes, I joined the others, like Greta Thunberg and thousands around the world. I walked out of school, with the blessing of Mum and the tight-lipped permission of my school. Although I know they are all 'proud' of me, they can't be seen to be outwardly encouraging civil disobedience. Mum stayed with me and brought a hot chocolate before I went back into school. I was frozen. Numb. But going back in with my signs was important. I needed to tell the other students why. Turning it over in my mind now, I wonder how effective it was. Was it just camaraderie? Were they only interested because I had rebelled? The feeling that I had to do something has been bottled up for years. And this single act has attracted more attention than all the other things I've done, all the work with raptors, the talks and awards for the things I've written. Is this more powerful? All the grown-ups are telling us how amazing this generation of activists is, commending our actions on social media or in the press, while doing what themselves? My generation has pulsed, and that is exciting. What doesn't sit well, though, is the search for 'leaders'. Climate leaders. Young leaders. The expectation seems preposterous. It appears that I'm now one of them. Just one single act of walking out and I've been crowned. It doesn't sit easy. It's not me, not me at all.

Sunday, 17 February

Last year, I saw my first frog towards the end of January. Not yet five degrees but there it was hopping across our path while we were hiking in Cuilcagh Mountain, perfectly content on the icy ground as it disappeared into the heather.

This morning, almost a month later than last year, I find one sheltering in bramble shadows, slinky skin and limbs tucked tightly, resting on mud and decaying oak leaves. I wait and wait for it to move, but the frog outdoes my patience and determination to stay motionless, because we're in a rush.

We've only stopped briefly at Peatlands Park, a nature reserve just off the M1, to break the journey back to Fermanagh, where we're heading for Granda Jim's birthday. He's seventy this year. It'll be so good to be with him and Nanny Pamela again – we haven't seen much of them since we moved east to County Down, and they're always so enthusiastic to spend time with us. Nanny is a couple of years older than Granda Jim, and has the energy of someone half her age. Granda has such sparkling eyes, and the kindest soul. Going west again feels halfway home and halfway to heartbreak.

The interlude at Peatlands Park is welcome, a stretching of our legs (with frog) before we travel the rest of the way. While we drive on, my thoughts wander to one of my first proper memories of being with Granda. It was when we were visiting the Crom Estate in Fermanagh. Lorcan wasn't born, yet the image is completely lucid: we're walking on a path adjacent to the ruined castle, which stands on the edge of a high bank overlooking Lough Erne. I drop down to listen for grasshoppers, but don't realise it's too cold to crouch down in the grass. I remember Granda's hand in mine as he told me about where he was born and how he walked miles to school every day. He told me how his father made saddles and school bags and delivered the post. I was mesmerised by his lilting voice, his gentle nature.

Mum thinks I invented this memory from a photograph, because I wasn't even two years old. But I'm convinced it's real. Maybe I processed more of it when I was older, attached new memories, but that moment left such a deep, warm feeling. I'm sure I was babbling on about something, probably starting with a 'Did you know'. I talked early, which was tough for everyone because I never ever stopped talking. Asking questions. Retelling facts about space or a woodlouse. Granda was so patient. He listened. And as we walked, the long grass tickled my legs. Usually, when I was out in local parks or playgrounds, I was taunted and mocked because of my longing to pass on information, to talk. It wasn't welcome. And it made me a target for bullying. There was none of this with Granda Jim. He listened, talked, and lifted me up in his arms to look at the castle. We felt the stone walls together, I kissed his head.

That day was one of my first memories and I hold it tenderly close. I saw the sadness in Granda's eyes, the way that Mum hugged him, her Daddy. Always Daddy. I can't remember stopping off to see his old cottage after the castle. But Mum has told me about the twisting and turning roads to Crieve Cross, and driving further on into the countryside until there it was, whitewashed, not much bigger than a tool shed; apparently I couldn't believe that so many people could fit inside. I still imagine the countryside around that cottage is perfect, with open skies and hawthorn everywhere.

I'm taller than Granda now, and when we arrive at the pub it's all hugs and hellos. I embrace him and Nanny extra tightly because life is fragile and achingly beautiful.

Sunday, 3 March

We live really close to the mountains: Commedagh, Donard and Bernagh dominate my everyday at school. It feels wonderful to be surrounded by them, and even better being able to whizz off on a whim towards them, like we're doing this morning because the persistent rain has subsided.

We're heading towards a car park on Slievenaman Road, so we can go for a quick traipse into Ott, just to shake off the lethargy left in our bones by the constant wet. As we climb, the air changes suddenly, and coming over the brow of a hill, we drive into a blizzard. We can't see two feet in front of the windscreen. It's unexpected and terrifying; we're lucky though because we can just about see the entrance to the car park.

This is the only snow I've experienced all winter, so we bale out of the car, not for a walk, but just to feel it. On our tongues and cheeks. Dimming all sound, snow creates so much space in the mind. Only in this weather can I process experience in real time with such clarity. Usually, it can be frightening because sights, sounds, feelings rush over me all at once. Sensory overloads that mean I can't properly process most of my experience until later in the day, in a dark room, when I relive the moment from scratch, spilling it all onto the page. In snow, things are different. Expansive thoughts unravel in the moment. There are fewer colours, less depth, less of everything. It really is quite a magical experience, secluded but with so much intense feeling, and even now in the howling wind and with cascading, blizzarding flakes of snow, my mind thrums differently. I can feel synapses sending signals. I can listen, I can hear. I can think and speak and feel and move all at once, instead of one process knocking clunkily into another. I never know if it makes sense to anyone else when I explain this feeling. I guess you would have to be me to really know. But I think we all have this sort of reaction to snow, just with different intensities.

The new palette of the land reveals bird tracks, and I suddenly remember being much smaller and close to the ground, following a fox track in the snow from our house across the road to Ormeau Park in Belfast. It was early on another Sunday, there was no traffic on the roads, no people, no sounds. Just fox tracks. Lorcan was in the sling because he was tired from not sleeping the night before and hadn't been walking long. We never found the fox itself, but it was following that mattered, a journey in the silence of the city, through one of the most peaceful days of my eight years living there. I'll never forget. I remember plunging my hand into the snow, to see what it would feel like, and then rolling over in it like a puppy wearing snow trousers. Laughing. Laughing with such relief.

From the car park on Slievenaman Road, I climb up some stone steps to a better viewpoint, the flakes dazzling, swirling, as my feet sink into the new depths. Everything is white except the stark outline of trees. I hold my face up to it, welcoming the tingle and taste. I want to stay for longer but Dad is anxious about our journey back. We have to go, just like that, and as we descend the hill, get into the car and drive away, the blizzard and the whiteness vanishes. All is as it was. A wet residue glistens on the land. There is no sign of snow. Did it actually happen? Did we all dream it? There's still snow on my boots and my hands are red raw, proof of Narnia. In and out of one beautiful, strange yet familiar world. Probably the last kiss of winter. I'm glad to have raised my head up to feel it.

Thursday, 21 March

There are unfurlings in the forest. Anemones and ferns are springing from patient earth, from dim and ancient spaces. Evensong is erupting and the airspace is once again crowded with music after the winter silence. Bluebells are on the cusp. Spring light and warmth are spreading across the mountains and into me. I have embraced the darkness but now this feeling of light is intoxicating, explosive and alive. By March I usually get impatient, itching for the spring. But not this time; I've been enchanted by every day, drinking in every moment.

Tomorrow I'll be taking to the streets of Belfast with other students, giving voice with many – not on my own like last time. I feel happier about it all. Civil disobedience is better in a group! And I won't have to bear so much weight and attract too much attention. The eco group will be taking a break soon, while I study for some of the GCSEs I'll be taking this year, but we've also been upping our game in school, giving up lunch breaks to gather round with banners, spreading awareness. I'm bursting with excitement about it all. I've never felt this way, it feels so alien, refreshing and electrifying. I wonder if it's because of the busyness. The action. The cramming in of wild experiences. I'm unfurling too, and feel so much more, dare I say it, stable. Not stagnant, never stagnant. And never blindly taking it all for granted. I think that would be disastrous. I know everything could change at any moment, but so many more pieces seem to be fitting together.

Last Sunday, for the festival of St Patrick, we made a pilgrimage to Glendalough, a glacial valley with two lakes and an ancient monastic settlement founded by St Kevin, my blackbird saint. It was the first time I'd been and all I wanted was to feel solitude and peace, but it was impossible. Standing on the bridge looking out over the restless Glendassan River, foaming over boulders, towards the round, thirty-foot tower, I could see people rushing here and there, tourists everywhere. But I was one, too. A pilgrim of St Kevin. And although it didn't feel like they were there for solace, as they swarmed with phones and clicking cameras and booming voices, rushing from one church to another, maybe all the people were longing for the same thing as me.

I was entranced by the place, with its granite structures covered in lichen and ferns, the walls with forests of polytrichum mosses and liverworts. We took our time, found pools of tadpoles, stopped to listen to full-throated mistle thrush singing in an overstorey of oak and understorey of holly, hazel and rowan, with bluebells, anemone and sorrel resplendent and glittering. The sun was shining, everything was golden and green, laden with morning rain, and I travelled inwards to fade out the voices and unnatural rumblings, tuned into the wildlife around me. Bláthnaid was in heaven too as she caressed the bark of the best climbing trees, and while resting her cheek on one mossy bough, she insisted she could hear a heartbeat. I could see it in her eyes, she really did feel it.

After circling the lower lake, we walked the longest of the Poulanass Waterfall routes, and by the time we reached Reefert Church we were by ourselves. Silence, as we climbed the steps of a rocky spur towards St Kevin's cell – only the foundations remain now, a circular set of jutting stones. There was still a granite slab, etched with the outline of downcast eyes, the noble nose, a slight smile. I really wasn't prepared for the intensity of emotion when I noticed a carved hand and a bird. A blackbird. I traced my finger over shapes in the shimmering quartz, and right there, under an overhang, there was a ladybird at rest. An orange ladybird, seeking shelter above the head of St Kevin.

My family carried on up towards the waterfall while I stayed there to rest my back against the stone. I looked out onto the lake, my body filling with shuddering – otter feelings. I thought of Kevin and his long journey from solitude to community, from being alone to being with others, and the way he must have found a space for both his own learning and the hospitality he offered to anyone who wanted it. I wondered how he balanced his need for silence with public work, and how his time with the elements and nature, with stone and wing, changed as more and more people came here.

I held out my hand to feel the tickle of the wind. A blackbird might never choose to nest and lay its eggs in my palm, but I know that my hand will always be outstretched, to nature and to people. Because we're not separate from nature. We are nature. And without a community, when you're always on your own, it's more difficult to share ideas and to grow. I'm so used to keeping my thoughts locked inside and being in a space where it's only me and my family. But now there are concentric circles, rippling out through a digital, online world into the very real world of activism, social action and interaction. It keeps on rippling. I have to drift and swirl with it, but always I'll need to retreat, back to the foundation stones of myself.

The vernal equinox has come and gone, and I'm now on the cusp of my fifteenth birthday, midway between late childhood and adulthood. Everything and nothing has changed. Again, Seamus Heaney's words are with me:

> Kevin feels the warm eggs, the small breast, the tucked
> Neat head and claws and finding himself linked
> Into the network of eternal life.

At equinox a couple of years ago, we visited Caldragh Graveyard on Boa Island in Fermanagh. A nestled place, tucked behind the lake shore, hugged by a circle of trees. Bluebells were everywhere, and some had been picked and placed on one of the early-Christian Janus figures, in the dipped hollow of its stone head. These almost 2,000-year-old figures both look forwards but in different directions, a duality. And that's what I felt on that day. I was thirteen, small in every way but with big thoughts. I put my hands on the stones and felt a rumbling ancestral roar. It was the sort of sound your mother might make if she were scolding you to warn that your life was in danger. Urgent. Pleading. I felt the heat of it when I placed my hand on my cheek.

On both Boa Island and Glendalough, with the traces of St Kevin, I felt gateways opening, choices to make, roads to travel. I have a longing to spend more time with the intricacies of nature, without the interactions and complications of people. I yearn for this simplicity, but I also want to go out into the world and weave my way, however overwhelming and painful it might be. Nature and us, at odds and at one.

As I ran to join my family for the last stretch of the walk at Glendalough, leaving St Kevin and the blackbird behind, a solar glare draped over us, connected us to the land with invisible strings. A longer, heavier line is about to be cast into the world. My heart is opening. I'm ready.

My soul is in the trees
It's in the sap that fills the wood
It's in the rings that tell her age
It's in the smoke that marks the days
It's in the fire in my heart
It's in the embers in the soot
It's in the place I put the ash
It's in the soil
It's in the grass
It's in the mouths of all the herd
It's in the beetles and the birds
It's in the feathers that I found one morning lying on the ground
It's hallelujah, aye and oh
It's where I've been and where I go
It's in the people that I meet
It's kneeling silent at their feet
It's ever dutifully yours
It stems my pride
And opens doors

from BOTTOM OF THE SEA BLUES *by Johnny Flynn*

Glossary

banshee (BAN-SHEE) ***bean sidhe***
Meaning 'fairy woman' in Irish. Banshees are considered harbingers of doom, calling out with blood-curdling screams. According to legend, if you came across a Banshee in the form of an old woman cleaning blood from clothes at a lakeside, this was a warning that you or a family member was about to die.

Beowulf (BAY-O-WOLF) ***Beowulf***
Considered one of the most important works of Old English literature, although the exact date of composition is unknown – the surviving manuscript dates from the late tenth to early eleventh century.

binn (BEN) ***binn***
Irish and Scottish Gaelic for mountain peak, particularly high ones. Often anglicised as Ben. The plural is Beanna. *See* also slieve.

Bláthnaid (BLAW-NID) ***Bláthnaid***
Means 'Blossoming One' in Irish, from 'Bláth' which is Irish for 'flower'.

Boa Island (BO ISLAND) ***Inis Badhbha*** (IN-IS BAA-V)
Badhbh's (Baa-v's) Island. Badhbh, meaning 'carrion crow', is the name of a Celtic war goddess. Boa Island in Lough Erne, is long and narrow, connected to the mainland by two bridges.

bodhrán (BOAR-ON) ***bodhrán***
An Irish frame drum made of goatskin commonly used in Irish traditional music.

bran (BRAAN) ***bran***
Bran (which means 'raven' in Irish), along with Sceolan, were Fionn Mac Cumhaill's legendary Irish wolfhounds. Their mother, Tuiren, had been transformed into a hound by a fairy woman.

cairn (KARN) ***carn***
A human-made pile of stones; in Ireland, often a prehistoric burial monument.

callows (KALLOWS) *caladh* (KALL'AH)
A river meadow. From the Irish 'caladh', this is a type of seasonally flooded grass wetland found in Ireland.

Caoimhín (KEE-VEEN) *Caoimhín*
Dara's third name (Kevin in English) and an important Irish saint of the sixth century who founded Glendalough Monastery (40 miles south-west of Dublin).

cashel (CASH-ELL) *cashel*
Means 'castle' in Irish but generally cashels are stone circular wall structures that date from the early Iron Age in Ireland.

Children of Lir (LEER) *Oidhe Chlainne Lir*
The tragedy of the Children of Lír. Lír, an Irish god and member of the Tuatha (TWO-UH-THA) De Danaan, married Aoife who turned his children from a previous marriage into (whooper) swans.

County Fermanagh (FUR-MAN-AH) *Fir (or Fear) Manach*
A county in the south-west of Northern Ireland, derived from the Irish 'Fir Manach' or Men of Manach. The most westerly county of Northern Ireland, it borders the Republic of Ireland and is one of the nine historical counties of the province of Ulster.

Country Park / Forest Park / Nature Reserve
In Northern Ireland there are seven government-owned Country Parks managed by the Northern Ireland Environment Agency (NIEA) including Castle Archdale Country Park alongside a number of nature reserves. There are also Forest Parks but these are managed by the Northern Ireland Forest Service.

Crocknafeola (CROCK-NA-FOAL-A) *Crock na feola*
'The Hill of Meat'. A small forest peak in the Mourne Mountains.

Cuilcagh Mountain (CULL-KEY) *Binn Chuilceach*
'Chalky Peak'. The mountain owes its name to the limestone geology of Fermanagh.

Dara (DA-RHA) *Dara / Dáire*
Meaning oak, wise, fruitful. Believed to be derived from the Irish Doire, the name is very common in Irish mythology.

Erne (ERR-N) *Éirne / Érann*
Name of a goddess who gave the name to the Érainn, an ethnic grouping widely scattered in Ireland and to whom belonged the Manaig of Fermanagh, a Celtic tribe originating from Belgium. The Erne river widens out into two large lakes: Upper and Lower Lough Erne.

Eimear (EE-MER) *Eimear*
Irish female name meaning 'swift'; the legendary wife of the Ulster hero warrior Cuchulain.

Enniskillen (ENNIS-KILL-IN) *Inis Ceithleann*
'Ceithleann's Island' Ceithleann was the wife of the legendary Formorian giant Balor. It is said that she swam for refuge to the island on which Enniskillen stands after inflicting fatal wounds on the king of the Tuatha Dé Danann at the battle of Moytura in Sligo.

Fianna (FEE-UH-NA) *Na Fianna*
Thought of as a special army of the High King of Ireland based in the ancient capital of Tara, County Meath.

Finn McCool (FIN-MAC-COOL) *Fionn Mac Cumhaill* (FEE-YUN MAC-COOL)
Chief of the Fianna and subject of many Irish legends.

Fomorian (FORE-MORE-IAN) *Fomhóire*
An evil race of beings who had their capital on Tory Island, County Donegal. They enslaved Ireland fighting battles with the Tuatha Dé Danann.

Glendalough (GLEN-DA-LOCH) *Glendalough*
'The valley of the two lakes', 40 miles south-west of Dublin in the Wicklow Mountains National Park. It is an ancient monastic city with monuments and round towers dating back to the sixth century, the early medieval period in Ireland. Founded by Saint Kevin.

goldfinch (LAA-SEER COLL-YEH) *lasair choille*
Translates into English as 'flame of the forest'.

inish (IN-ISH) *inis*
Irish for island, with variants of *inch* or *inse*. It is often anglicised as Inish or Ennis, i.e. Enniskillen, or *Inis Ceithleann* (Kathleen's Island).

Isle of Inishglora (IN-ISH GLOR-RA) *Inis Gluaire*
An uninhabited island off the west coast of Ireland beside the Mullet Peninsula, Erris, County Mayo. Dara's Great Grandmother was born in the Erris area.

lagan (LAG-HAN) *an lagáin*
Means river of the low lying district. The River Lagan is the main river that flows through Belfast City – the Lagan towpath is a beautiful walking and cycling path through woods on the outskirts of Belfast. Originally called 'laogh' meaning calf.

lon dubh (LAWN DOO / DUV) *lon dubh*
Blackbird in Irish.

Lorcan (LOR-CAN) *Lorcan*
Means 'the fierce one' in Irish.

lough (LOCK) *loch*
Lough is the anglicised or English version of the Irish Gaelic word loch, meaning lake. The use of lough tends to be restricted to Ireland and does not extend to anglicised Scottish place names.

Lough Derravaragh (LOCH DERRA-VAR-OCH) *Loch Dairbhreach*
The Children of Lír spent three hundred years here, before moving to the straits of the Moyle between Ireland and Scotland and then three hundred more in the west of Ireland between Erris in Mayo and Inishglora.

Loughnabrickboy (LOCK-NA-BRICK-BOY) *Loch na breac buí*
'Lough of the yellow trout', located in Big Dog Forest, Fermanagh.

Mallacht (MALL-OCT) *Mallacht*
The witch whom Fionn Mac Cumhail and his hounds chased across Fermanagh. Her name means 'cursing one'. She stopped during the chase to turn the hounds into stone hills known as Big Dog and Little Dog (Bran and Sceolan).

McAnulty (MAC-A-NULL-TEE) *Mac An Ultaigh*
Meaning 'Son of an Ulsterman'. This clan is a sect of the Mac Donleavys who ruled the Kingdom of Ulster, or Ulaid, from Downpatrick until it was conquered by the Norman knight Sir John de Courcy in 1177.

Mourne Mountains (MOURN) *Múrna / Beanna Boirche*
A granite mountain range in south County Down named after the Irish clann Múghdhorna or in modern Irish, Múrna, who settled there in the 1300s. They are also known as the Mountains of Mourne made famous in a song written by Percy French in 1896 and covered by many artists including Don McLean. The more ancient name of Beanna Boirche (banna-bor-ka) is said by some to mean peaks or benns of the mystical herdsman Boirche who tended the King of Ulster's cattle in the third century.

Quoile (QU-OIL) *An Caol*
Meaning 'the narrow', the River Quoile is a river in Downpatrick, County Down. On the north bank sits Inch Abbey, a pre-Norman Celtic monastic settlement. Quoile nature reserve is situated on both sides of the river.

Róisín (ROW-SHEEN) *Róisín*
An Irish name meaning 'little rose'.

Samhain (SAH-WIN) *Samhain*
Gaelic festival with pagan roots, marking the end of the harvest and the start of the darker winter time. Historically observed in Scotland, Ireland and the Isle of Man, it was traditionally celebrated from 31 October to 1 November, and Christianised as

Hallowe'en.

Sceolan (SH-KYO-LAN) *Sceolan*
Brother of Bran and one of Fionn Mac Cumhaill's legendary Irish wolfhounds.

scréachóg (SCRA-OH-G RAIL-YA-GA) *scréachóg reilige*
Irish for Barn Owl, meaning the graveyard screecher.

Sea of Moyle (MOY-ULL) *Sruth na Maoile*
The strait of sea that separates South West Scotland and Northern Ireland, also known as the North Channel or Irish Channel. It is possible to see across in clear weather. The narrowest point is approximately 12 miles wide. Maoile is Gaelic for Mull (of Kintyre) meaning bald summit.

slieve (SLEEVE) *sliabh*
There are many words in Irish to describe mountains, most commonly sliabh, anglicised as slieve, and found in the names of Irish mountains or ranges; also used for hills. *See* also binn.

Slieve Donard (SLEEVE DONN-ARD) *Sliabh Dónairt*
Meaning Dónairt's mountain, it is nearly 2,800 feet high (850 m), impressively rising from the sea and one of the twelve chief mountains of Ireland. It is the tallest mountain in Northern Ireland. Saint Dónairt, formerly a local pagan king and warrior, became a follower of St Patrick and lived there as a hermit.

Slieve Muck (SLEEVE MUK) *Sliabh Muc*
One of the Mourne Mountains. Meaning the 'mountain of the pig' or 'mountain of wild boar'. The northern slope is the source of the River Bann, the longest river in Northern Ireland.

Slievenaslat (SLEEVE-NA-SHLAT) *Sliabh na slat*
Located in Castlewellan Forest Park, meaning mountain of the rods or sticks – there's still a lot of willow and hazel scrub here, perhaps once used for weaving, basket-making, etc.

Stormont (STOR-MONT) *Stormont*
The building in Belfast that houses the Northern Ireland devolved Government Assembly and executive, set up after the 1998 Good Friday Agreement.

Tamnaharry (TAM-NA-HARRY) *Tamhnach an Choirthe*
The 'clearing in the upland of the standing stone'. Tamnaharry, near Mayobridge, Newry, County Down, has a significant standing stone, an ancient megalithic structure, on the hill looking down on it. The farm in Tamnaharry is where Dara's great-grandfather James McAnulty was raised.

uaigneas (OO-IG-NUSS) *uaigneas*
Not easy to translate into English but can mean 'a sense of loneliness, an eeriness'.

Acknowledgements

With heartfelt gratitude to my family, for your unswerving, unconditional love and support. For giving me wings and allowing me to fly in my own direction, to the beat of my own drum. Your patience, sacrifice, humour and adventurous spirit have allowed me to thrive and soar. I hope I can one day repay you all – Mum, Dad, Lorcan, Bláthnaid and Rosie. You're the best!

To Adrian at Little Toller, for not trying to 'adult' my voice in the editing process. For originally giving me the opportunity to tell my story, in all its irreverent rawness and childish wonder. Sincere thanks to Gracie, Jon and Graham.

Heartfelt thanks to all the team at Ebury; especially Hana, Drummond, Jo, Caroline and Anna for this beautiful new edition. Your belief that my book could fly into many more hands and hearts has been humbling and uplifting.

To my agent Cathryn, Jess and all at Curtis Brown for supporting me and my family. For navigating us through this strange world of publishing.

To Tony Smith, my amazing scout leader and friend, who showed me that I can push my limits, fall out of my comfort zone and try 'hard things' – and then celebrate success! Our wild scout camps are amongst my best childhood memories; eating wood sorrel atop a wooded quarry cliff, canoeing, rambling – these experiences made me. And although these memories aren't mentioned in the book, they are one of the main reasons it exists.

Dr Eimear Rooney and Dr Kendrew Colhoun, ornithologists extraordinaire, your guidance and expertise hasn't turned off my raptor obsession – so you're stuck with me. Sorry!

To Chris Packham, for your friendship and patience, while being my teenage-angst soundboard. You watered my roots and gave me the confidence to grow. For your unswerving devotion to the natural world and for elevating the voices of all young naturalists and activists. (I'll stop now before you hate me!)

To Robert Macfarlane, for the hagstone, the literary advice and stalwart support, enthusiasm and encouragement. From the very beginning, you championed my words and my voice. You are a gentleman and a scholar (that's the highest form of praise on the island of Ireland).

My school friends and community – you have turned my world on a positive axis. I may be constantly spinning out of control, but your gravity constantly levels my spirit.

To these organisations for providing me with scaffolding to climb and shout loud for the natural world: Northern Ireland Raptor Study Group, Ulster Wildlife (The Grassroots Challenge), Royal Society for the Protection of Birds, #IWILL campaign, Our Bright Future and The National Trust. I hope we can continue to work together for a better world – you all have my continued support and energy.

And to nature: my source, roots, beat and thrust. My canopy. My shield and sword.

D. McA.

County Down, 2021

About the Author

Dara McAnulty is a 17-year-old autistic naturalist, conservationist and activist from Northern Ireland. After writing his online blog 'Naturalist Dara' for over three years, and articles for many UK Wildlife NGOs, Dara wrote his first book, *Diary of a Young Naturalist*. It won the Wainwright Prize for Nature Writing – Dara the youngest ever winner of a major UK literary prize – as well as the Books Are My Bag Non-Fiction Readers Award, Newcomer of the Year at the An Post Irish Book Awards, HAY Festival Book of the Year, and the Big Issue Book of the Year. Dara is a passionate and fervent campaigner for the natural world and a dedicated fundraiser, volunteer and wildlife recorder. He is the youngest ever winner of the Royal Society for the Protection of Birds Medal for services to conservation and nature. He is also the recipient of 10 Downing Street's 'Points of Light' and the winner of the *Daily Mirror* Young Animal Hero award. He lives with his family and Rosie the rescue-greyhound at the foot of the Mourne Mountains in County Down.